Asia in Transition

Volume 7

This book series is an initiative in conjunction with Springer under the auspices of the Universiti Brunei Darussalam—Institute of Asian Studies (http://ias.ubd.edu.bn/). It addresses the interplay of local, national, regional and global influences in Southeast, South and East Asia and the processes of translation and exchange across boundaries and borders. The series explores a variety of disciplinary and interdisciplinary perspectives.

Kwen Fee Lian (Ed.), *Multiculturalism, Migration, and the Politics of Identity in* Singapore, 2016, 978-981-287-675-1 (Asia in Transition 1)

Kwen Fee Lian, Md Mizanur Rahman & Yabit bin Alas (Eds.), *International Migration in Southeast Asia: Continuities and Discontinuities*, 2016, 978-981-287-711-6 (Asia in Transition 2)

Mikio Oishi (Ed.), *Contemporary Conflicts in Southeast Asia: Towards a New ASEAN Way of Conflict Management*, 2016, 978-981-10-0040-9 (Asia in Transition 3)

Victor T. King, Zawawi Ibrahim & Noor Hasharina Hassan (Eds.), *Borneo Studies in History, Society and Culture*, 2017, 978-981-10-0671-5 (Asia in Transition 4)

More information about this series at http://www.springer.com/series/13611

Mario Ivan Lopez · Jafar Suryomenggolo
Editors

Environmental Resources Use and Challenges in Contemporary Southeast Asia

Tropical Ecosystems in Transition

ubd INSTITUTE OF ASIAN STUDIES

Springer

Editors
Mario Ivan Lopez
Kyoto University
Kyoto
Japan

Jafar Suryomenggolo
National Graduate Institute
for Policy Studies
Tokyo
Japan

ISSN 2364-8252 ISSN 2364-8260 (electronic)
Asia in Transition
ISBN 978-981-10-8880-3 ISBN 978-981-10-8881-0 (eBook)
https://doi.org/10.1007/978-981-10-8881-0

Library of Congress Control Number: 2018935222

Printed on acid-free paper

This Springer imprint is published by the registered company Springer Nature Singapore Pte Ltd.
part of Springer Nature
The registered company address is: 152 Beach Road, #21-01/04 Gateway East, Singapore 189721, Singapore

Acknowledgements

This volume arose out of the editor's interest to engage in a more multidisciplinary dialogue with researchers who work in Southeast Asia and talk across disciplinary boundaries, rather than within them. Between 2011 and 2016, the Center for Southeast Asian Studies (CSEAS), Kyoto University, ran a Large-Scale Research Program "Promoting the Study of Sustainable Humanosphere in Southeast Asia." One part of this program promoted research on tropical biomass with a focus on Southeast Asia to clarify the circulation and use of resources within the region. Both editors are trained in the humanities and social sciences, yet worked side by side with natural scientists, ecologists as well as other researchers from disciplines more familiar to them. Out of exchanges, seminars and workshops with our colleagues, rose the idea and concept for this volume, particularly the aim to foster greater dialogue between researchers who straddled the social/natural science divide.

CSEAS has a long history of working within and across the natural sciences, social sciences, and humanities and this volume sits upon the intersections that cut across these three areas. Research at CSEAS has been characterized by a commitment to long-term fieldwork which has created a substantial body of scholarship through interdisciplinary discussion. This volume arises out of an urgent need to engage with different disciplines to forge new approaches toward understanding Southeast Asia.

We are indebted to our fellow colleagues who were on this program, in particular the former Director of CSEAS, Kono Yasuyuki who was the Secretariat head of the above project. We would also like to offer our thanks to colleagues who looked over this manuscript to offer suggestions for improvement and in particular, Nathan Badenoch (CSEAS), who gave us insightful advice for our introduction. We are also grateful to have been invited to publish with Universiti Brunei Darussalam-Institute of Asian Studies. In 2015, CSEAS held a seminar with UBD-Institute of Asian Studies and the Southern Institute of Social Sciences (SISS) in Ho Chi Minh,

Vietnam, and this led to an invitation from Associate Professor Jérémy Jammes to publish in the IAS-Asia in Transition Series. We hope that this volume will expose researchers to the rich empirical studies that highlight the transitions and transformations that presently sweep over Southeast Asia.

Kyoto, Japan
Tokyo, Japan
April 2018

Mario Ivan Lopez
Jafar Suryomenggolo

Contents

Editors and Contributors

About the Editors

Mario Ivan Lopez is Associate Professor at the Centre for Southeast Asia Studies, Kyoto University. He received his PhD from Kyushu University and has published papers on migration in Southeast Asia in *Global Networks* and *Philippine Studies.* He is currently a member of a Large-Scale Research Program running at the Centre that is promoting the study of a Sustainable Humanosphere in Southeast Asia (2011–16). E-mail: marioivanlopez@cseas.kyoto-u.ac.jp.

Jafar Suryomenggolo is Assistant Professor at National Graduate Institute for Policy Studies (GRIPS), Tokyo, Japan. He is graduated from University of Indonesia (Law, 2000) and Kyoto University (Ph.D., Southeast Asian Studies, 2010). His research interests are on working-class politics and political changes in contemporary Southeast Asia. E-mail: j-suryomenggolo@grips.ac.jp.

Contributors

Nyein Chan Department of Pollution Control and Waste Management, University of Forestry and Environmental Science, Nay Pi Taw, Myanmar

Panuwan Chantawannakul Department of Biology, Faculty of Science, Chiang Mai University, Chiang Mai, Thailand

Tran Van Do Research Institute for Sustainable Humanosphere (RISH), Kyoto University, Kyoto, Japan; Silviculture Research Institute, Vietnamese Academy of Forest Sciences, Hanoi, Vietnam

Haris Gunawan Peatland Restoration Agency of Republic of Indonesia, Jakarta, Indonesia

Kazuyuki Iwasa Faculty of Humanities and Social Sciences, Kochi University, Kochi, Japan

Samai Jai-In Alternative Energy Foundation, Institute of Thailand (AEF), Bangkok, Thailand

Muanpong Juntopas Climate Change Adaptation Specialist, Asian Development Bank (ADB), Bangkok, Thailand

Yuka Kiguchi Mekong Watch, Tokyo, Japan

Yasuyuki Kono Center for Southeast Asian Studies (CSEAS), Kyoto University, Kyoto, Japan

Mario Ivan Lopez Center for Southeast Asian Studies (CSEAS), Kyoto University, Kyoto, Japan

M. D. Macabuhay Department of Environmental Science, Ateneo de Manila University, Quezon City, Philippines

Bao Maohong History Department, Peking University, Beijing, China

I. A. Navarrete Department of Environmental Science, Ateneo de Manila University, Quezon City, Philippines

D. P. Peque Department of Forest Science, Visayas State University, Baybay, Philippines

Takahiro Sato Faculty of Agriculture and Life Sciences, Hirosaki University, Hirosaki, Japan

Tamotsu Sato Department of Forest Vegetation, Forestry and Forest Products Research Institute, Tsukuba, Japan

Sabrina Shaw International Institute for Sustainable Development (IISD), Buenos Aires, Argentina

Jafar Suryomenggolo National Graduate Institute for Policy Studies (GRIPS), Tokyo, Japan

Shinya Takeda Graduate School of Asian and African Studies (ASAFAS), Kyoto University, Kyoto, Japan

Shinsuke Tomita Graduate School of Environmental Studies, Nagoya University, Nagoya, Japan

Kazuo Watanabe Center for Southeast Asian Studies (CSEAS), Kyoto University, Kyoto, Japan

Lamphoune Xayvongsa Faculty of Forest Science, National University of Laos, Laos, Lao People's Democratic Republic

Le Zhang School of Geography and Environment, Jiangxi Normal University, Nanchang, China

Abbreviations

ACMES	Ayeyawady-Chao Phraya-Mekong Economic Cooperation Strategy
ACNNR	Agreement on Conservation of Nature and Natural Resources
ADB	Asian Development Bank
AEDP	Alternative Energy Development Plan
AFTA	ASEAN Free Trade Area
AGB	Above Ground Mass
ANSOFT	Asian Network for Sustainable Organic Farming Technology
AOP	Assembly of the Poor
ASEAN	Association of Southeast Asian Nations
BAPPENAS	*Badan Perencanaan Pembangunan Nasional* (Indonesian National Planning Board)
BAR	Philippine Bureau of Agricultural Research
BHD	Bio Hydrofined Diesel
BNPB	*Badan Nasional Pengendalian Bencana* (Indonesian National Board for Disaster Management)
BPS	*Badan Pusat Statistik* (Indonesian Central Bureau of Statistics)
BTL	Biomass to Liquid
CCD	Colony Collapse Disorder
CPO	Crude Palm Oil
CRB	Coarse Root Biomass
CVM	Contingent Valuation Method
DBH	Diameter at Breast Height
DEDE	Thai Department of Alternative Energy and Efficiency
DENR	Philippines Department of Environment and Natural Resources
EDC	*Electricite du Cambodge* (Cambodia Electricity Company)
FAO	Food and Agricultural Organization
FELDA	Malaysian Federal Land Development Authority
FMB	Philippines Forest Management Bureau
GDP	Gross Domestic Product
GHG	Green House Gas Emissions

GMS	Greater Mekong Subregion
GSP	General System of Preferences
HANPP	Human Appropriation of Net Primary Production
IAASTD	International Assessment of Agricultural Knowledge, Science and Technology for Development
IFOAM	International Federation of Organic Agriculture Movements
IMF	International Monetary Fund
IPCC	Intergovernmental Panel on Climate Change
IRRI	International Rice Research Institute
JICA	Japanese International Cooperation Authority
LAP	Laotian Land Allocation Program
LFA	Land and Forest Allocation
LIFE	European Commission's Financial Instrument for the Environment
LIRE	Lao Institute for Renewable Energy
MPOA	Malaysian Palm Oil Association
MPOB	Malaysian Palm Oil Board
MRC	Mekong River Commission
MRP	Mega Rice Project
MSSRC	Mekong Sub-Region Social Research Center
NAC	New Agricultural Countries
NAFOSTED	Vietnamese National Foundation for Science and Technology Development
NAOP	Philippines National Organic Agriculture Program
NCR	Native Customary Right
NEP	Net Ecosystem Production
NFP	Philippine National Forestation Program
NGP	Philippine National Greening Program
NOAB	Philippine National Organic Agriculture Board
NPP	Net Primary Production
NPV	Net Present Value
NTFP	Non-Timber Forest Product
OA	Organic Agriculture
OM	Organic Matter
OPTA	Organic Producers Trade Association
PAFO	Laotian Provincial Agriculture and Forestry Office
PEMANDU	Malaysian Performance Management and Delivery Unit
PGS	Participatory Guarantee System
PNB	*Permodalan Nasional Berhad* (Malaysian National Fund Management Company)
PORAM	Palm Oil Refiner's Association of Malaysia
PPO	Peach Palm Oil
REDD+	Reducing Emissions from Deforestation and Forest Degradation or Enhancement of Carbon Stocks
RIWP	Root Impermeable Water Sheet
RSPO	Roundtable on Sustainable Palm Oil

UNCTAD	United Nations Conference on Trade and Development
UNEP	United Nations Environment Program
UNESCAP	United National Economic and Social Program for Asia and the Pacific

List of Figures

List of Tables

Chapter 1
Introduction

Environmental Resources Use and Challenges in Contemporary Southeast Asia

Mario Ivan Lopez and Jafar Suryomenggolo

Abstract Southeast Asia's rich environment has been slowly eroded as the region taps its rich resources pursuing a growth-oriented paradigm. It is confronting serious issues such as the decrease of biodiversity and tropical forests and the degradation of peatlands. Concurrently, social issues tied to energy-dependent growth are never far from people's lived experiences in the region, and have intensified over the last two decades. The purpose of this edited volume is to introduce dynamic approaches to the study of Southeast Asia's environmental diversity from different disciplinary perspectives mainly through a natural/social science interface. It brings together scholars whose research is on the region's environmental resource use and shared ecological challenges under the effects of present day globalization to offer insights for possible future directions.

Keywords Southeast Asia · Environmental degradation · Tropics
Resources use · Socio-ecology · Agro-ecology · Development

Introduction

Southeast Asia is one of the world's most diverse regions in terms of human societies and its environment. Home to over 600 million people, it is rich in ethnic, religious and cultural composition. An increasingly important economic hub, the region, until recently, maintained a level of biodiversity not found in other parts of the developing world. It has come to be characterized by the flow of people, goods, money, and information, especially with the establishment of a common market along the lines of an ASEAN Economic Community (AEC). Indeed, the nature

M. I. Lopez (✉)
Center for Southeast Asian Studies (CSEAS), Kyoto University, Kyoto, Japan
e-mail: marioivanlopez@cseas.kyoto-u.ac.jp

J. Suryomenggolo
National Graduate Institute for Policy Studies (GRIPS), Tokyo, Japan
e-mail: j-suryomenggolo@grips.ac.jp

M. Lopez and J. Suryomenggolo (eds.), *Environmental Resources Use and Challenges in Contemporary Southeast Asia*, Asia in Transition 7,
https://doi.org/10.1007/978-981-10-8881-0_1

of Southeast Asia's transformation, in terms of its diverse societies and environmental biodiversity, at both a regional and global level, represents both internal integration and far reaching and deeper connections into the modern world system. The result of changes brought about by this has meant that its biodiversity has been rapidly transformed as the region taps its rich energy resources and biomass. The diversity of conditions in the region has, over the centuries, defined, shaped, and informed human–environment interactions, livelihood, and economic practices and sustained agro-systems of varying sizes. The unique nature of the region with good alluvial plains has been characterized by monsoons where reliable rainfall distribution has allowed for the continual existence of high-density populations. However, Southeast Asia's diverse environment has undergone vast transformations in the recent years, the result of political-economic integration into the global world system through export-oriented industries predicated upon the twin paradigms of growth and development.

Over the course of the late twentieth century, intensive flows of investment in agriculture, new technological regimes, and forms of management have been deployed intensifying cultivation and production to support increasingly urbanized populations in Southeast Asia. These forms of investment have developed an energy-dependent growth paradigm with a high level of path dependency. Subsequently, this has fostered new patterns of consumption reorganizing social life in the region. Scholars have closely detailed the agrarian and rural transitions of the past 30 years as nations, at different speeds and intensities, pursued development projects and numerous studies have critically dealt with their practice and implementation within the region (See Rigg 2012, 2015; Dixon and Drakakis-Smith 1997). These studies are important reference points for this volume for they have recognized the changes wrought upon the region and critically shed light on how development practices have impacted its ascendency in Asia.

As processes of development have transformed the region, new infrastructure linkages, predicated upon strong conceptualizations of modernization, have connected Southeast Asia to other regions allowing for natural resources to flow out. This has been a boon for some nations, producing patches of prosperity while others have lagged behind. The integration that has been witnessed has been criticized as being based upon a neoliberal capitalist model of economic growth leading to a "poverty of sustainable development" (Rigg 2015). Economic growth oriented forms of development (economic expansion) that have been introduced to the region are, in part, responsible for the stark reality it now faces. It is confronted by a steep decrease in biodiversity and tropical forests (Yamada 1997; Fujita and Samejima 2016; Hughes 2017); the degradation of peatlands (Miettinen et al. 2012a); the creation of vast oil palm plantations (Pye and Bhattacharya 2013); and resources dependence on minerals extraction, timber resources, and plantation crops (Coxhead 2007). This has led some to call this a "pathology of habitat loss" (Sodhi and Brook 2011), one linked to modern anthropogenic changes. These forms of development have come to define discussions over how best to respond to region-wide environmental changes and what forms of mechanisms to develop to manage/govern these.

ASEAN and the Formation of an Environmental Regime

One response to the above has been through the formation of institutional mechanisms. When ASEAN was formed in 1967, environmental issues were not considered a primary concern. It was only later and since 1977, alongside a development-oriented agenda, that environmental issues were given consideration. Over the past few decades, this regional institution has come to recognize shared local and transboundary environmental issues such as haze pollution, marine fisheries degradation, and soil degradation to name a few (Nguitragool 2011; Bush and Marschke 2016). These have come to take on an important point of departure through the creation of agreements that tackle national and regional level changes in the environment, climate change and its impact on the region's biodiversity. These agreements fall into a broader global discussion taking place, yet the institutional frameworks, legal provisions, strategies, programs, and roadmaps all point to a complex fragmented socioeconomic landscape which includes both countries with some of the highest GDP in the region as well as some of the lowest. Since the mid-1980s, various institutional arrangements and agreements were launched with a focus on the principle of state's responsibilities toward the region's environment.

One of the core ASEAN principles that has informed discussions is the taken-for-granted ASEAN "way" of noninterference and consensus-based decision-making. However, adherence to this principle has meant that discussions have often lagged behind the dynamic contexts and conditions of a rapidly transforming shared environment and ecosystems which have been penetrated and transformed by the expansion of human habitats and market forces. This has led to calls for reenvisioning the way issues should be responded to and the development of real strong intra-regional political will and strong institutions that can engage transboundary issues that threaten human security. Regional integration has made remarkable in-roads through the development of manufacturing and production bases around the region and although there has been much progress in alleviating poverty, this has been accompanied by a continued increase in environmental degradation and resources extraction that has yet to be tackled through the ASEAN Charter. Nevertheless, forms of governance predicated upon a more holistic approach to the region are starting to bring about what some have observed as "environmental regionalism" (Elliot 2012a) where states acknowledge the benefits of cooperation within the limits of ASEAN principles.

As a regional institution, ASEAN has increasingly expanded its institutional capacities and remit to deal with environmental problems in the region. Koh and Karim (2016) note that it has developed mechanisms, policies, and agreements to tackle national and shared environmental issues such as the Manila Declaration (1981) and the Agreement on Conservation of Nature and Natural Resources (ACNNR) (1985) acknowledging a shared environment. In terms of transboundary issues that have impacted nations, the Kuala Lumpur Accord of Environment and Development (1990) made clear nation's shared duties and responsibilities toward each other. ASEAN Summits have also come to serve as important platforms for nations to recognize transboundary pollution issues.[1] These agreements are, in a sense, a sign of the rise of regional environmental governance tools but have been criticized for falling into the ASEAN way trap: nonbinding, governed by state sovereignty and non-interventionist.[2]

Nevertheless, over the past decade, we are also witnessing a turning point in ASEAN's attitudes and strategies to environmental issues where it has become more active working on environmental governance in the region. To strengthen cooperation among member countries, ASEAN has made serious efforts to balance economic growth and environmental sustainability through various programs. Its commitment to environmental protection and conservation, however, is still hampered by the lack of political will of its member states to execute and implement the agendas that they have agreed upon. This is further exacerbated by the fact that in reality, economic development still takes precedence over a legally binding regional environmental regime (Koh and Karim 2016). In this regard, there is room for the region's national policy-makers to take up this issue to bring a change in their respective countries through political discourse in the public sphere.

Aiming Beyond Development

As previously mentioned, one of the guiding practices and goals over the past 30 years has been to pursue variegated forms of development and this has wrought unprecedented changes: massive anthropogenic habitat modification, overexploitation of wildlife, the loss of lowland forests (in particular primary growth forest), the degradation of water basins and peatlands. These are all presenting clear threats to biodiversity (Sodhi et al. 2004; Fisher et al. 2011; Elliot 2012a). This has not been without international recognition of the region's vital role as a part of the tropics in regulating biospheric and geospheric processes. This can be seen in the deployment of financial instruments related to carbon sinks or the global mechanism REDD+ (Reducing Emissions from Deforestation and Forest Degradation or enhancement of carbon stocks) to reduce (and dissuade) the capture and transformation of resources; from forest biomass, agricultural products, and biomaterials for regional consumption and global export. Nevertheless, these have not substantially slowed down the development of intensive region spanning agro-industrial production systems. These include large-scale plantations for cash cropping (Nevins and Peluso 2008), high value commodities that link regions, peoples and products

[1]In 1995, the ASEAN plan of Transboundary Pollution was adopted and 2002 marked the beginning of a push to reduce haze pollution in Southeast Asia with the ASEAN Agreement on Transboundary Haze Pollution (2002), finally ratified by 10 countries in 2014. In a way issues, such as transboundary pollution, acted as catalysts that forced some nations to the table, tackling regional inertia especially when public health became an issue.

[2]See Heilmann (2015) for detailed critique and Koh and Karim (2016) for overview of the different programs, their phases and the environmental protectionist regime that have developed since 1977.

(Tsing 2015), and the large-scale conversion of forests to food crops with subsequent deforestation (Wilcove et al. 2013; Fisher et al. 2011).

Another crucial issue the region face is the governance of social-ecological systems (SESs) the politics of scale in concern to resources, their extraction, and allocation. State bureaucracies, government agencies, and public planners can decisively influence the speed, flow, intensity, and scale of (as well restrictions) the extraction use and allocation of resources within, between, and across nations. Where, for example, community-based conservation has been implemented to regulate economic behavior, market-oriented forms of agrarian change have further rationalized resources as part of a commodity landscape emphasizing the difficulties of balancing centralized/decentralization forms of governance. With both regional and internal demands for resources on the increase, there has never been a more crucial time for governance that is informed and can make decisions based upon the reconfigurations taking place (Wolfram and Roth 2011). This is one of the challenges that researchers faces in putting across empirical data driven results that can impact upon policy-makers and effect change.

In relation to these above concerns, this volume aims to illustrate empirical and dynamic approaches to the study of Southeast Asia from combined disciplinary perspectives. It brings together scholars who work across Southeast Asia, ecologists, agronomists, social and political scientists whose concerns cut across the classic divide that exists in research. It concentrates on field-based studies through which scholars have developed basic, yet comprehensive research projects to contribute to discussions in three broad areas: geospheric circulations, ecosystems and their economic significance, and the politics of energy and resources use. All researchers in this volume use empirical fieldwork-oriented methods which include both social science and natural science methodologies employed to ask questions and elucidate responses toward their subject matter.

It introduces field-driven research to link the social and natural sciences, both methodologically and empirically, to present current multidisciplinary approaches to the study of Southeast Asia's environmental changes and resource management under the effects of present day globalization within the region. Government policy-makers are often unaware of the research results and knowledge accumulation on these important subjects, and their short-term policies tend to be carried out without deeper deliberation based on hard research. This volume hopes to alert those who are in a position to inform policy-makers on environmental issues and possible frameworks of action and open up.

In light of these overlapping arenas of engagement—the social, economic, and ecological—the kind of methodologies and research topics researchers use will decide the information we produce and how we can utilize this to examine the multidimensional driving forces of change in Southeast Asia. New research approaches are duly needed to better understand and articulate what conditions influence transformations in the region; human communities' impacts on local fauna and flora; degradation of terrestrial and aquatic biomes; regional politico-economic policy; and the subsequent influxes of populations into rapidly urbanizing and expanding cities. This volume arises out of a concerned response to these issues and how they interact with the complexities of Southeast Asia. It provides a snapshot of issues from the level of the community through to the national and transnational to provide an overview of the conditions and transformations that are shaping the seamless web of interactions that define socioeconomic activities in the region vis-à-vis its ecosystems.

Changing Socio-Ecological Landscape(s) in Southeast Asia

Scholars have repeatedly pointed out that Southeast Asia's socio-ecological landscapes have been and continue to be at a crucial juncture point (King 1998; Bradshaw et al. 2009; Bickford et al. 2012). One of the main issues is the increasing resource use for consumption of energy to fuel the region's urban growth. Over the past 30 years, numerous development projects were conceived, planned, and implemented, yet there was little concern to integrate environmental issues into national programs of economic development, beyond natural resources extraction and exploitation. The last decade may have seen a shift through some attempts to include environmental planning and management, but this has had little impact on the aspirations toward economic progress nor forced policy-makers to seriously consider alternative paradigms of economic development. The resulting—and continuing—intensification of developing Southeast Asia has led to an inevitable strong impact on the overall environment.

To place the discussion into a broader global context, the twentieth century also witnessed the doubling of human appropriation of net primary production (HANPP) to influence nearly three-quarters of all vegetated land (Krausmann et al. 2013). The socioeconomic impacts of one species on global NPP are a cause for severe concern (Haberl et al. 2007). At present, human societies appropriate roughly 20% of terrestrial NPP globally and east and south-central Asia appropriate 72% of their regional NPP (Imhoff and Bounoua 2006). Additionally, almost 11% of the Earth's land surface is now allocated for crop production but uses 70% of water drawn from aquifers, streams, and lakes (FAO 2011). Urban expansion across many regions, but in particular, Southeast Asia will impact upon its most productive croplands: by 2030 large losses will take place (as high as 80%) in Asia and Africa, with Asia experiencing the highest percentage loss (Bren d'Amour et al. 2016).

Southeast Asia is one of the world's richest biomes, but experiences some of the highest proportional rates of rainforest loss and is currently experiencing one of the highest rates of deforestation in the world (Sodhi et al. 2010). Different levels of dependency on land, the conversion of forested areas to agricultural land, the creation of oil palm plantations, commercial logging, mining and hydropower dams are playing out at different speeds and intensities in the region. As numerous societies in Southeast Asia transition from agrarian socio-ecological regimes to highly urbanized ones, we will see greater pressure placed upon livestock populations, cereal production, growth in rural populations, and the expansion of agricultural areas as per capita GDP rises (Ashraf et al. 2017). At present, just under

60% of Southeast Asians are now living in dense highly urbanized cities[3] (Singru 2015) and this is set to rise in the coming decades. These urban developments will be made all the more acute by the formation of new middle classes accompanied by new levels of affluence and consumption which will place pressure on productive lands. This expansion will not be solely limited to the large urban megalopolises that are coming to define the region.

Most countries in the region will witness the emergence of new smaller cities further encroaching on and transforming surrounding hinterlands. Through the expansion of urban areas, coastal mangroves have undergone transformation for aquaculture and oil palm use (Richards and Friess 2016). Tropical peatlands, covering an estimated 440,000 km^2 ($\sim$10% of global peatland area) (Page et al. 2004), have been consistent carbon sinks throughout human history, but these are now coming under intense land-use pressure (Miettinen et al. 2012a, b). Since 1990, over 54,000 km^2 was deforested in Peninsular Malaysia, Sumatra and Borneo alone (Miettinen et al. 2012a). 56 million hectares of arable land are now experiencing different levels of decline in soil fertility, an issue the region now faces as a whole (Elliot 2012b). Marine Fisheries in the region, in particular those in Indonesia, Vietnam, Cambodia and the Philippines will all suffer negative impacts as rapid economic development and growing populations consumption of fish places pressure on poorly regulated aquaculture industries (Barange et al. 2014).

Contemporary Challenges and Insights

Qualitatively understanding the immense changes re-shaping Southeast Asia poses a series of daunting questions with political implications. If environmental degradation, deforestation, and transnational agribusiness activities can destabilize the ecological foundations that support life in the region, how can empirical observations and knowledge claims translate into action? Through what means and channels can scientific findings and the application of knowledge specific to the region shape political decisions? These questions relate squarely to what Turner calls "a politics of knowledge," one that shapes outcomes in the arena of environment politics (Turner 2011). How can research such as that in this collection contextualize the specificities of the region, especially in terms of its ecology and how peoples, institutions, businesses, and governments co-interact, to produce applied knowledge that arises out of engagement with the material world?

Earlier work has already emphasized the need for political ecology approaches toward locating resources use in the tropical forests of Asia, taking the political as a primary point of analysis (Lye et al. 2003). Political ecology, here, is defined as the critical analysis of problems of natural resource appropriation and the

[3]Most cities in the region are also vulnerable to at least one type of natural disaster (UNDP ESA 2016).

political-economic origins of resource degradation (Lye et al. 2003, p. 4). Other researchers (Rambo 1983) have situated the dynamic relationships between social and ecological systems focusing on how the former respond to choices available and normalize behavior, norms, and practices, including, those that shape extractive political practices that institutionalize themselves. Yet, for all these analyses and frameworks, we have not been able to slow down or ameliorate the projected biotic extinctions that will occur as a result of human activities in the region (Sodhi et al. 2010). This has led for calls from ecologists for the integration of concerted region-wide conservation planning, including a gender-equitable approach to natural resources governance (Colfer et al. 2016), and technology transfer between high-income to low-income countries to push for change that engages with social issues (see Sodhi et al. 2010). More recently, other researchers have taken country level researchers toward climate change and the vulnerabilities nations will have to confront in the near future (see Bruun and Casse for their detailed overview of Vietnam 2013). This volume is mainly constituted by field researchers who respond to these issues from their current respective research topics to offer an overview of issues the region is facing as climate change begins to impact upon the region's present and future.

In spite of the negative prognosis for the region's future, researchers have also noted that there exists a complex interplay between different institutions and the strong political role they can play in reorienting both perception and transition toward different forms of land utilization. Truong et al. (2017), show how, in the case of Vietnamese farmers driven reforestation, smallholder agricultural intensification pathways can lead to reforestation of less productive hillsides where positive economic conditions exist. The region is constituted by unique ecological contours and historically a "rural" topography has been at the heart of the region.[4] Acknowledging the role of farmers as agents with unique agro-ecological knowledge suited to rehabilitating degraded lands should hold a place in broader discussions that relate to caring for the region's agro-ecosystems. However, vast infrastructural change and new economic opportunities linking to urban hubs have led to a spatial relocation of populations. This has produced new structural conditions and the integration of the region into the world system through rural industrialization, and in many cases, the displacement of localized knowledge. This has meant that a new infrastructure is densely meshing not just the region, but creating strong transport, commercial and economic linkages to meet the demands of markets in China, East Asia and the world (Nevins and Peluso 2008; Murray 2014; Tsing 2015). This is forming new commodity chain links and intensifying the extraction of resources.

A common thread that ties the authors in this volume who work across the social science/natural science domain is a dual approach they take. As researchers,

---

[4]See Rigg (2001) for a critical overview of the way the "rural" in Southeast Asia has been politicized by categorizing and homogenizing it as distinct from the urban in terms of economy and society. This has led to a reductionist and static view of the region's diverse rural hinterlands.

they approach the environment as a space(s) where different geographies and overlapping ecologies coexist at both a national and regional level. They also approach these spaces as research sites which are not free from discursive creation and bound to transformations wrought by human interactions on local ecosystems. They are aware of the fact that the "environment" is a social construction, one with a specific historical discursive trajectory that is used very broadly and with different interpretations in political debates at national, regional and global levels. All authors, through microlevel approaches to their field sites, are aware of the debates and stakes at hand in defining a region; the struggle of claiming and producing economic value that can help develop nations on one hand, and on the other, the consequences that arise from short-term economic and political policies. Climate change, subsequent transformation of the environment, and the impact upon diverse ecosystems are a regional concern for Southeast Asian nations. This is acknowledged by the development of a framework for "environmental policy" (at a community, national and regional level, including in ASEAN as noted above) through civic education reform and the institutionalization of laws which have, to date been fragmentary. We acknowledge these dimensions which form a backdrop across the chapters in this volume.

Another framework that underlies some concerns in this volume is environmental governance and the growing awareness of how local communities play an important role in claiming a say in how governance impacts upon their livelihood. Over the past 20 years, this has come to shape and dominate discussions on how natural resources are utilized. Governance, in the case of Southeast Asia, has been a "practice that entails the blurring of law in societal and political modes of rule-setting and enactment" (Boer et al. 2016, p. 86). It is an array of practices where different actors, communities, and stakeholders aim to influence how resources are equitably managed within the framework of livelihood security and social justice. These latter two are of the utmost importance where some common resources in the region are transboundary and effect a large number of people who are dependent upon them.

The Mekong Delta, often conceived and portrayed as an entity, is a prime example of a shared transboundary resource—a contested waterscape (Molle et al. 2010; Lazarus et al. 2011) where competing demands, claims, access, institutions, and transnational entities (such as the Asian Development Bank (ADB), International Monetary Fund (IMF) and the Mekong River Commission (MRC)) all operate to constitute a complex regulatory regime. Institutions are embodied through political-economic relationships, themselves embedded in both local, regional, and transnational decision-making structures as exemplified by the MRC. Whereas the technocratic economic management of resources dominated the programs and legal frameworks and provisions in the 1970s and 1980s, we now see resources rationalized by a plethora of actors, many of whom live off the very contested resources that sustain their livelihood. What is now at stake is how local knowledge embedded within systems, and often centuries old, can be re-attuned to deal with the new realities of appraising, managing, and fostering resources in a pragmatic and equitable fashion. In a way, there has been a move away from centralized national management to decentralized forms of environmental governance, but successes through this approach have been limited.[5] What is required are detailed case studies that can contribute to a meaningful discussion on the importance of research-based praxis on Southeast Asia's ecosystems and how they are to be governed. As we will show below, this volume is structured to reflect this objective.

Structure of the Volume

This volume is organized within three broad interrelated areas. The first area deals with geospheric circulations, ecosystems, and the region's diverse agro-energy systems. In this part, we present two case studies. One that deals with the importance of Southeast Asia's biosphere and geosphere in the tropics for sustainable development, and a second that focuses on bio-energy procurement within the Mekong region. The second area focuses on Southeast Asia's eco-resources on the revitalization of forestry and economic transitions as a result of changing land use. It consists of five original case studies on the critical issues of carbon certificate and reforestation, and also on human–nature interactions. The third area focuses on the issues of politics of energy and resources use. It consists of one study on the rise of organic agriculture in the context of economic development, one case study on state interventions in dam development, and one final study that examines government policies toward swidden cultivation and subsistence livelihood.

As discussed in the previous sections, Southeast Asia is defined by its complex socio-ecological systems. To speak meaningfully of the region requires an understanding of the complexity at play not just within a multilayered system, but also an observation, appraisal, and critique of the logics that operate within it. Kono et al. (Chap. 1) examine the role of human behavior in agricultural and water use in mainland Southeast Asia and examine what kinds of mechanisms have formed in human societies that have developed in tandem with the biosphere and geosphere of the tropics. The chapter shows that there coexist layered development approaches with different timescales and types of care specific to the local geosphere and biosphere. Crucially, they strongly argue for the importance of experientially fostered knowledge to be integrated with scientific knowledge to cope with the diversity and complexity of the tropics and to observe the integrality of social structures and the practices that arose from them in response to the specificities of the environment. The chapter raises the question as to how different development paths can be produced from within the ecological contours of regions with the own unique human–nature interactions. The industrial revolution of the past 150 years and the formation of modern nation-states allowed human societies to organize and

[5]Decentralization and devolution in the region have had limited success in part due to the absence of democracy, transparency, and accountability.

acquire new technological regimes which have reorganized social life on an unprecedented scale. Twentieth-century development unleashed a great production capacity that has supported materially abundant lifestyles yet, human activities have led to a cascade of critical environmental changes producing a negative feedback loop transforming both the biosphere and geosphere. Western societies and their technologies, along with scientific knowledge coupled with capitalist economic growth oriented production, have extended all over the Earth. These technologies come with an array of practices, policies, discourses, and political consequences that have rationalized the environment as an infinite source of resources that can fuel economic growth ad infinitum. This chapter, implicitly questioning the modernity discourse as it has played out in Southeast Asia, highlights the need to think about an alternative economic mode of production that is attuned to the conditions of the region.

Shaw et al. (Chap. 2) scrutinize the policy spaces needed to develop the foundations for a sustainable agriculture while fueling transformation in the Mekong region. This chapter, based on fieldwork with those working in agro-energy industries, highlights a need for a paradigm shift in the food–energy–water nexus to move toward greater policy integration. The use of bioenergy may promote energy security, revitalize rural economies, and reduce greenhouse gas emissions if implemented at a local community level. However, meeting food and energy needs in the Mekong in the twenty-first century, especially in Southeast Asia, will require a range of approaches to shift the current development paradigm in order to enhance natural resource governance and climate-friendly agricultural practices. This chapter highlights that if current practices prevail, Thailand, the focus of the study is likely to use an unsustainable model of trade-led development at a regional level. Evidence indicates that synergies will be difficult to capture in the current policy climate that separates people's wellbeing, social progress, and eco-system sustainability from intensifying resource use in the Mekong. To this end, the authors put forward an alternative development strategy to build on synergies between small-scale initiatives and agro-energy sustainability.

Natural resources management practices and vulnerability have been the focus of recent studies in Vietnam (von Platen-Hallermund and Thorsen 2013) in the context of the social vulnerability of communities. Considering the economic viability of REDD+ at a community level, Tran and Tamotsu (Chap. 3) discuss Net Ecosystem Production (NEP) in the tropical broadleaved forests of Copia Natural Reserve, northwest Vietnam, and highlight the need for a carbon certificate to support local communities. In recent years, ecologists have focused on estimating NEP to understand the role of forests against increasing concentrations of CO_2 in the atmosphere, a high priority issue of concern. Research in this chapter asserts the importance of using forest carbon certificates so that forest protectors (local communities) can raise their bargaining power to improve income and protect forests in a sustainable fashion. Importantly, the chapter raises discussion on carbon prices and how they should be identified locally based on economic and natural status.

Navarrete et al. (Chap. 4) focus on the role of soil information in supporting the analysis of reforestation schemes in the Philippines. Forests support not only ecosystem processes, but also social and economic services for human survival. This chapter details the deforestation and degradation and the vast annual forest cover that is lost in the Philippines. Over the last two decades, the Philippine government has devoted a considerable amount of resources to both the rehabilitation and reforestation of degraded forest land. While large-scale reforestation projects have been initiated, most replanting efforts have been far from successful. The major contributing factors to this failure were, and still are, a lack of understanding of the nature and characteristics of the soil, particularly site quality evaluation and a lack of site-species matching. This chapter argues that the rational use of forest resources should be based upon accurate knowledge of soil properties. The authors of this chapter argue that the soil is the single most important factor affecting the survival, growth, and development of trees and thus, influences the success of reforestation programs. They offer a detailed analysis of soil factors that limit the success of these reforestation projects and present current research and development efforts to reforest degraded forest land, particularly the National Greening Program (NGP), and the role of the reforestation program on soil carbon sequestration.

Gunawan (Chap. 5) provide a case study from Riau and West Kalimantan and offer a template for how to rehabilitate degraded peatland at a local community level. Peat swamp forest ecosystems are now recognized for the crucial various environmental services they provide locally, nationally, and globally. However, under increasing land scarcity, the pressure to use them for agriculture has exponentially increased. In the case of Indonesia, small-scale local systems have shown to be more adaptive to peat swamp environments, rendering minimum environmental degradation, but large-scale conversion and usage, especially that which involves fire for land clearing and canal drainage has led to an abrupt decline in peat environment quality over the past 30 years. Since the peat fires of 1997, much research has evaluated peatland environmental functions in economic terms and has found that in some cases the economic benefits from their environmental functions far exceed profits from agricultural activities. However, for landholders, conserving peat swamp forests translates into forgoing benefits due to relatively high net present value or net profits that can be earned if land is converted to agricultural use. If environmental services are public goods that are not marketable regardless of their level of economic value, then those goods that farmers produce from agriculture are marketable with tangible benefits for them. This chapter shows that the opportunity cost of CO_2 emission reductions by conserving peat swamp forests from conversion to oil palm is far higher than the current registered emission reduction compensation price. The chapter suggests that opportunity costs are higher than the carbon market price, yet a carbon market is not yet available, especially for peat forest conservation. The author call for urgent development strategies to establish viable compensation alternatives to landholders beyond the carbon market and provide a roadmap for rehabilitation that can benefit both communities and the peatlands that they live from.

Much recent research has focused on commodity products such as oil palm to show what kinds of highly specific supply chains have arisen to deal with the demand for commodities, in turn further creating dense extractive agro-industrial linkages. These have intensified massive plantation development projects through transnational ventures that have extended beyond Southeast Asia to link into others. Iwasa (Chap. 6) focuses on the palm oil industry in Malaysia. This chapter examines and contextualizes the structure of the industry as a set of agro-industrial production linkages which stretch into both Malaysia and outside of it as part of a transnational commodity chain. Second, it emphasizes the transnationalization of agribusiness capital and the very hard impacts this will have in the short and long term for patterns of investment and the development trajectory of the palm oil industry in Malaysia. The country was, since the 1960s, a pioneer in developing plantations to become a leading palm oil producer/exporter and, alongside Indonesia, came to rapidly convert its tropical ecosystems over to a monocultured landscape. This chapter shows how competition between Malaysia and Indonesia has led to the transnational expansion of agribusiness capital. This is further exacerbating environmental disruption and land grabbing by Malaysia in Indonesia as well as in West Africa in response to an increased demand for palm oil in other regional markets.

As landscapes have been transformed, other species have come under pressure from the socioeconomic demands and imperatives to consume both within and outside the region. Panuwan Chantawannakul, a Thai apiologist, (Chap. 7), focuses on honey bees (*Apis.*) to situate the history of beekeeping, beekeeping practices, and their introduction into the region in the twentieth century. This chapter offers an overview of the impact of changing environments as well as human activities on honey bee populations in Southeast Asia. Southeast Asia is a major biodiversity hotspot for bees who are one of the key pollinators in the region. They are known for the crucial role they maintain for plant biodiversity in forests and agricultural crops for human food production. In Southeast Asia, traditional knowledge and beekeeping practices to harvest bee products have long been embedded in local cultures. Nevertheless, the significant transformations in the region have severely affected wild and domesticated honey bee populations. Under these conditions, some species have adapted to their new environments, yet other populations have been rapidly decreasing. The combination of human activities—land use and deforestation, newly emerging diseases, and climate change means that they are struggling to adapt and maintain species survival in the region.

The above chapters offer clear discussions on how to think about the impact of governance, ecological concerns, and environmental programs and how they impact upon livelihood. These important issues are tightly linked to the fact that the region has been experiencing unprecedented socioeconomic changes, in terms of market expansion, a rise in consumption, and energy demands over the past decade. Although these changes are more apparent among those living in urban settings, communities far from capital cities are challenged to improvise adaptive strategies, especially through their livelihood and agricultural practices—despite the limited capital they may possess.

In one context, there has been phenomenal growth in the demand for organic products for consumption, especially among the middle class in the region. Bao Maohong (Chap. 8) examines the rise of organic agriculture (OA) and its latest development in the Philippines. OA has been a rising emergent market and has been slowly integrating into the Philippines national economy. The 1980s was marked as a decade of unsustainable economic activity with negative impacts on the country's environment leading to some to proclaim the "green revolution" a failure. Out of this arose a combinational force of the international agriculture movement and the reinvention of traditional Philippine agriculture leading to the creation of OA in the Philippines. This was notable through its recognition by a republic Act in 2010 which put this new form of agriculture on the national agenda through the efforts of pro-OA NGOs and government bodies. This chapter highlights how OA could be an alternative to input-intensive organic agriculture however, any mainstreaming and integration will depend on the reconstruction of socioeconomic relations with the complex political system of the Philippines.

Another series of developments that are changing the region's waterways is the contentious intensification of dam building projects across and within nations in their rush to hydropower generation. At the heart of the push to secure the development of energy sources is, as Smits makes clear, two key themes: ideas of what constitutes modernity and sustainability (Smits 2015). The former plays a crucial role in defining how state/corporate actors recognize energy as a source to transform livelihood and develop national agendas, yet the material consequences of conceptualizations of development play out in the enactment of governance, management regimes, and its impact on local communities. This is crucial in the knowledge production "industry" of the Mekong where conflicting regimes of governance exert force over this large transboundary river (Rajesh et al. 2013). Damming across the region has, as such, often implicated concrete governance issues and its impact on community livelihood. Kiguchi (Chap. 9) focuses on the latter which is often affected after the construction of dams along Mekong River tributaries. It focuses closely on the case study of Pak Mun Dam in Northeastern Thailand to show how local communities have been left to bear the negative impacts of dam construction, and how these affected both fisheries and degraded local knowledge. The Pak Mun Dam case serves as a departure point to understand the socio-environmental changes that will arise with the construction of a new dam in the region, the Lower Sesan 2 Dam in Cambodia. The chapter shows how local communities have responded and coped with socio-environmental changes wrought by dam construction in the region and provides sobering lessons for ongoing damming projects which are transforming the "waterscapes" of the region (Molle et al. 2010). What Kiguchi's work emphasizes is how local knowledge is not only degraded, but sidelined by notions of development. Ultimately, the energy trajectory pursued has a dual impact upon both community and the local ecosystem.

The livelihood of many Southeast Asian's, especially those that live in the highlands, has often been challenged by transformations in the forms of agricultural activities for subsistence and government policies toward those who live off the land. People's lives in the highlands have often been caught up in externally imposed driving forces that shape both the agricultural and socioeconomic environment (Neef and Ekasingh 2007). Nyein et al. (Chap. 10) offer a detailed case study on swidden cultivators in northern Laos and evaluate the potential impacts of a Land and Forest Allocation (LFA) program that was initiated in Laos in the 1990s to stabilize swidden agriculture practices. Laos, as with other countries in the region, has seen some of its more traditional forms of agriculture displaced by other forms of agro-economic activities. This chapter looks at the importance of involving communities in government planning that effects them, especially as now the government of Laos has been active in promoting the internationally recognized and sanctioned mechanism of Reducing Emissions from Deforestation and Forest Degradation (REDD+) to enhance forest stocks. What this chapter calls for is a deeper integrated understanding of the socioeconomic conditions of people who carry out and live from swidden cultivation for subsistence and cash income, and how it can be integrated at a local and regional level.

Despite the diverse topics discussed across the chapters in this volume, to varying degrees, most of the authors share some common points. They are all concerned with the changes that are transforming the region. They are concerned about supportive measures for local people engaged in subsistence or reliant upon ecosystem services (Chap. 10); including communities in the future management of peatlands (Chap. 5) or forest ecosystems (Chaps. 3 and 4); and wanting to find ways to respect the inherent appropriateness of agricultural paths of development attuned to the tropics in the region (Chap. 1). All authors agree informing policy-makers, actors, organizations, and peoples and their communities must be based upon empirical perspectives that arise from Southeast Asia's diverse ecological reality.

In summary, the case studies presented here provide an interesting comparative lens for scholars to look into what is taking place at the moment in the environment in the region, the pace of its change, and the efforts of researchers investigating real challenges. Despite all of these, the case studies in this volume offer some paths for rethinking how we can engage with the ecological priorities the region faces. Paths include conceptual thinking, alternative perspectives, and policy recommendations to respond to major problems. Most of the contributors in this volume are presenting viable proposals for change—with some degree of pessimistic caution and understanding of limitations. These chapters are meant to raise our awareness on the shared ecological challenges the region faces, and alert us to the directions we must move into meet the looming crises. Environmental NGO advocates and policy-makers in the region could also learn some new insights from the discussions to formulate strategic plans for action. As many different factors interact in these overlapping issues, this volume wishes to reiterate that cooperation among stakeholders is key to avoiding a "tragedy of the commons" and erring on the side of caution (Bradshaw et al. 2009). Whether Southeast Asia can avoid irreversible degradation that weakens the foundations of human society, other species, and the environment will largely depend on the different synergies that arise between researchers, policy-makers, government organizations and agencies, and most importantly civil society movements. Southeast Asia is no longer at a crossroads and the alliances formed now will decide how the region faces ongoing challenges in the twenty-first century.

References

Asharf, J., et al. (2017). Assessment of bio-physical, social and economic drivers for forest transition in Asia-Pacific region. *Forest Policy and Economics, 76,* 35–44. Retrieved February 15, 2017, from http://dx.doi.org/10.1016/j.forpol.2016.07.008.

Barange, M., Merino, G., Blanchard, J. L., Scholtens, J., Harle, J., Allison, E. H., et al. (2014). Impacts of climate change on marine ecosystems production in societies dependent on fisheries. *Nature Climate Change, 4,* 211–216. https://doi.org/10.1038/nclimate2119.

Bickford, D., Poo, S., & Posa, M. R. C. (2012). Southeast Asian biodiversity crisis. In D. J. Gower, K. G. Johnson, & J. E. Richardson (Eds.), *Biotic Evolution and Environmental Change in Southeast Asia* (The systematics association special volume 82). Cambridge: Cambridge University Press.

Boer, B., Hirsch, P., Johns, F., Saul, B., & Scurrah, N. (2016). *The Mekong: A socio-legal approach to river basin development.* London: Earthscan.

Bradshaw, C. J. A., Sodhi, N. S., & Brook, B. W. (2009). Tropical turmoil: A biodiversity tragedy in progress. *Frontiers in Ecology and the Environment, 7,* 79–87.

Bren d'Amour, C., et al. (2016). Future urban land expansion and implications for global croplands. *PNAS.* (early edition). Retrieved February 10, 2017, from www.pnas.org/cgi/doi/10.1073/pnas.1606036114.

Bruun, O., & Casse, T. (Eds.). (2013). *On the frontiers of climate and environmental change: Vulnerabilities and adaptations in central Vietnam.* Hiedelberg: Springer.

Bush, S. R., & Marschke, M. (2016). Social and political ecology of fisheries and aquaculture in Southeast Asia. In P. Hirsch (Ed.), *The Routledge handbook of the environment in Southeast Asia.* London: Routledge.

Colfer, C. J. P., Basnett, B. S., & Elias, M. (Eds.). (2016). *Gender and Forests: Climate change, tenure, value chanins and emerging issues.* Oxon: Routledge.

Coxhead, I. (2007). A new resource curse? Impacts of China's boom on comparative advantage and resource dependence in Southeast Asia. *World Development, 35*(7), 1099–1119.

Dixon, C., & Drakakis-Smith, D. (1997). *Uneven development in Southeast Asia.* Aldershot: Ashgate.

Elliott, L. (2012a). ASEAN and environmental governance: Strategies of regionalism in Southeast Asia. *Global Environmental Politics, 12*(3), 38–57.

Elliot, L. (2012b). Shades of green in east asia: The impact of financial crises on the environment. In S. Breslin (Ed.), *East asia and the global crisis.* London: Routledge.

Fisher, B., et al. (2011). The high costs of conserving Southeast Asia's lowland rainforests. *Frontiers in Ecology and the Environment, 9*(6), 329–334.

Food and Agriculture Organization of the United Nations (FAO). (2011). *The State of the world's land and water resources for food and agriculture: Managing systems at risk.* Abingdon: Earthscan.

Fujita, M., & Samejima, H. (2016). The biodiversity of Southeast Asian tropical rainforests. In K. Mizuno, M. S. Fujita, & S. Kawai (Eds.), *Catastrophe and regeneration in Indonesia's peatlands: Ecology, economy and society.* Kyoto-CSEAS Series on Asian Studies 15. Singapore: NUS Press.

Haberl, H., et al. (2007). Quantifying and mapping the human appropriation of net primary production in earth's terrestrial ecosystems. *PNAS, 104*(31), 12942–12947.

Heilmann, D. (2015). After Indonesia's ratification: The ASEAN agreement on transboundary haze pollution and its effectiveness as a regional environmental governance tool. *Journal of Current Southeast Asian Affair, 34*(3), 95–121.

Hughes, A. C. (2017). Understanding the drivers of Southeast Asian biodiversity loss. *Ecosphere, 8*(1), 1–33.

Imhoff, M. L., & Bounoua, L. (2006). Exploring global patterns of net primary production carbon supply and demand using satellite observations and statistical data. *Journal of Geophysical Research, 111,* D22S12. https://doi.org/10.1029/2006jd007377.

King, V. T. (Ed.). (1998). *Environmental challenges in Southeast Asia*. London: Routledge.
Koh, K.-L., & Karim, Md. S. (2016). The role of ASEAN in shaping regional environmental protection. In P. Hirsch (Ed.), *The Routledge handbook of the environment in Southeast Asia*, pp. 315–333. London: Routledge.
Krausmann, F., et al. (2013). Global human appropriation of net primary production doubled in the 20th century. *PNAS, 100*(25), 10324–10329.
Lazarus, K., Badenoch, N., Dao, N., & Resurreccion, B. P. (2011). *Water rights and social justice in the Mekong region*. London: Earthscan.
Lye, T.-P., de Jong, W., & Ken-ichi, A. (Eds.). (2003). *The political ecology of tropical forests in Southeast Asia: Historical perspectives*. Kyoto Area Studies on Asia, CSEAS, Vol. 6. Kyoto: Kyoto University Press.
Miettinen, J., Shi, C., & Liew, S. C. (2012a). Two decades of destruction in Southeast Asia's peat swamp forests. *Frontiers in Ecology and the Environment, 10*(3), 124–128. https://doi.org/10.1890/100236.
Miettinen, J., Hooijer, A., Shi, C., Tollenaar, D., Vernimmen, R., Liew, S. C., et al. (2012b). Extent of industrial plantations on Southeast Asian peatlands in 2010 with analysis of historical expansion and future projections. *GCB Bioenergy, 4*(6), 908–918.
Molle, F., Foran, T., & Käkönen, M. (Eds.). (2010). *Contested waterscapes in the Mekong region: Hydropower, livelihoods and governance*. London: Earthscan.
Murray Li, T. (2014). *Land's end: Capitalist relations on an indigenous frontier*. Durham: Duke University Press.
Neef, A., & Ekasingh, B. (2007). Institutional framework for sustainable land use. In F. Heidhues, L. Herrmann, A. Neef, S. Neidhart, J. Pape, P. Sruamsiri, D. C. Thu, & A. Valle Zárate (Eds.), *Sustainable land use in mountainous regions of Southeast Asia: Meeting the challenges of ecological, socio-economic and cultural diversity*. Berlin: Springer.
Nevins, J., & Peluso, N. L. (Eds.). (2008). *Taking Southeast Asia to Market: Commodities, nature and people in the neoliberal age*. Ithaca: Cornell University Press.
Nguitragool, P. (2011). *Environmental cooperation in Southeast Asia: ASEAN's regime for transboundary haze pollution*. Routledge Contemporary Southeast Asia Series. Abingdon: Routledge.
Page, S., Wüst, R. A. J., Weiss, D., Reley, J. O., Shotyk, W., & Limin, S. H. (2004). A record of Late Pleistocene and Holocene carbon accumulation and climate change from an equatorial peat bog (Kalimantan, Indonesia): Implications for past, present and future carbon dynamics. *Journal of Quaternary Science, 19*(7), 625–635. https://doi.org/10.1002/jqs.884.
Pye, O., & Bhattacharya, J. (Eds.). (2013). *The palm oil controversy in Southeast Asia: A transnational perspective*. Institute of Southeast Asian Studies (ISEAS): Singapore.
Rajesh, D., Lebel, L., & Manoram, K. (Eds.). (2013). *Governing the Mekong: Engaging in the politics of knowledge*. Selangor: SIRD.
Rambo, T. A. (1983). *Conceptual approaches to human ecology*. Research Report No 14. Honolulu, East-West Center.
Richards, D., & Friess, D. A. (2016). Rates and drivers of mangrove deforestation in Southeast Asia 2000–2012. *PNAS, 113*(2), 344–349.
Rigg, J. (2001). *More than the soil: Rural change in Southeast Asia*. London: Routledge.
Rigg, J. (2012). *Unplanned development: Tracking change in South East Asia*. London: Zed Books.
Rigg, J. (2015). *Challenging Southeast Asian development: The shadows of success*. London: Routledge.
Singru, R. N. (2015). *Regional balanced urbanization for inclusive cities development: Urban-rural poverty linkages in secondary cities development in Southeast Asia*. No. 11. ADB Southeast Asia working paper series. Retrieved February 15, 2017, from https://www.adb.org/sites/default/files/publication/161353/sewp-11.pdf.
Smits, M. (2015). *Southeast Asia energy transitions: Between modernity and sustainability*. Surrey: Ashgate.

Sodhi, N. S., & Brook, B. W. (2011). *Southeast asian biodiversity in crisis*. Cambridge: Cambridge University Press.
Sodhi, N. S., et al. (2004). Southeast Asian biodiversity: An impending disaster. *Trends in Ecology & Evolution, 19*(12), 654–660.
Sodhi, N. S., et al. (2010). The state and conservation of Southeast Asian biodiversity. *Biodiversity Conservation, 19*(2), 317–328.
Truong, D. M., et al. (2017). Forest transition in Vietnam: A case study of northern mountain region. *Forest Policy and Economics, 76,* 72–80.
Tsing, A. L. (2015). *The mushroom at the end of the world: On the possibility of life in capitalist ruins*. Princeton: Princeton University Press.
Turner, M. D. (2011). *Knowing nature: Conversations at the intersection of political ecology and science studies*. Chicago: University of Chicago Press.
United Nations. (2016). *The World's cities in 2016: Data booklet.* Economic and Social Affairs (ESA). Retrieved February 15, 2017, from http://www.un.org/en/development/desa/population/publications/pdf/urbanization/the_worlds_cities_in_2016_data_booklet.pdf.
von Platen-Hallermund, T., & Thorsen, A. M. (2013). Natural resource management impact on vulnerability in relation to climate change: A case in a micro-scale Vietnamese context. In O. Bruun & T. Casse (Eds.), *On the frontiers of climate and environmental change: Vulnerabilities and adaptations in central Vietnam*. Hiedelberg: Springer.
Wilcove, D. S., et al. (2013). Navjot's nightmare revisited: Logging, agriculture, and biodiversity in Southeast Asia. *Trends in Ecology & Evolution, 28*(9), 531–540.
Wolfram, D., & Roth, R. (2011). The good, the bad, and the contradictory: Neoliberal conservation governance in rural Southeast Asia. *World Development, 39*(5), 851–862.
Yamada, I. (1997). *Tropical rain forests of Southeast Asia: A forest ecologist's view*. (Trans: Peter Hawkes). Honolulu: Kyoto University and University of Hawai'i Press.

Part I
Geospheric Circulations

Chapter 2
Reconsidering Development Mechanisms of Tropical Agriculture: Focusing on Micro-Development in Mainland Southeast Asia

Yasuyuki Kono, Takahiro Sato, Kazuo Watanabe, Shinsuke Tomita and Le Zhang

Abstract This chapter looks at everyday human behavior, agricultural development, and water use in mainland Southeast Asia and asks what kinds of mechanisms have formed in human societies that have developed in tandem with the biosphere and geosphere of the tropics. With the industrial revolution and the formation of modern nation-states, human societies acquired new technological regimes leading to the reorganization of social life on an unprecedented scale. In the twentieth century, this unleashed a great production capacity that has supported materially abundant lifestyles. Yet simultaneously, this production capacity triggered various environmental issues and subsequent measures to deal with them. Currently, the tropics are a particularly striking example of the rapid decline and loss of species diversity and the irreversible environmental degradation brought about by subsequent large-scale agricultural development. This chapter aims to show that there coexist various development approaches with different timescales and different types of care for a local geosphere and biosphere, and argues the importance of experientially fostered knowledge to be integrated with scientific knowledge to cope the diversity and complexity of the tropics in Southeast Asia.

Y. Kono (✉) · K. Watanabe
Center for Southeast Asian Studies (CSEAS), Kyoto University, Kyoto, Japan
e-mail: kono@cseas.kyoto-u.ac.jp

T. Sato
Faculty of Agriculture and Life Sciences, Hirosaki University, Hirosaki, Japan

S. Tomita
Graduate School of Environmental Studies, Nagoya University, Nagoya, Japan

L. Zhang
School of Geography and Environment, Jiangxi Normal University, Nanchang, China

M. Lopez and J. Suryomenggolo (eds.), *Environmental Resources Use and Challenges in Contemporary Southeast Asia*, Asia in Transition 7,
https://doi.org/10.1007/978-981-10-8881-0_2

Keywords Input-intensive agriculture · Rainfed paddy cultivation
Micro-development · Farm-level water management · Chinese market
Commercial crop · Experiential knowledge

Introduction

Over the last several decades, tropical agriculture has undergone drastic development. In 1961, the population of the tropics stood at 1.4 billion, yet by 2013 it increased 3.1-fold to 4.2 billion (United Nations Population Division). Land area under cultivation expanded by 42% from 563 million ha in 1961 to 700 million ha in 1994, and reached 798 million ha in 2013 (FAO), and a majority of individuals comprising the additional population are engaged in agriculture for livelihood. Over the same period, in the temperate world, the population increased only 1.7-fold, and cultivated land decreased 5% (FAO). For tropical agriculture, the latter half of the twentieth century was a period of predominant challenge for achieving increases in production to meet a surge in food demand resulting from an explosive growth of population and upgrading living standards. This trend has been further accelerated in the twenty-first century due to intensive global flows of agricultural investment and production.

Within this unprecedented demographic transition, attempts have been made to adopt an approach "packaging" advanced technology for use in agricultural development. Major components of this technology have been large-scale irrigation and drainage systems, the deployment of agricultural machinery, chemical fertilizers and other chemicals, and improved crop cultivars. The adoption of this packaged approach was widely triggered in Southeast Asia by the Green Revolution, which disseminated a rice cultivation system of high-yielding cultivars by means of controlling the production environment coupled with the intensive use of chemical fertilizers (Hayami and Kikuchi 2000). Without a doubt, such innovations have mitigated the (re)occurrence of food crises and served as a foundation for economic growth in tropical regions.

However, it is also true that input-intensive agriculture has caused a wide range of unexpected negative impacts on both the physical and biological environments or, in other words, the regional geosphere and biosphere. Technology-based input-intensive agriculture has also drastically affected soil structure, soil microorganisms, destabilized farmland ecosystems, and triggered soil erosion (McNeill 2000; Kyuma 2005; Fröhlich et al. 2013). Furthermore, the control of regional water circulation heavily influences local weather and climate systems and these anthropogenic activities have produced a cascade of changes in local flora and fauna as well as in both soil and hydrologic environments (Kozan 2010). These effects, induced by input-intensive agriculture, have spread and accumulated in the interconnected web formed between the geosphere and biosphere (Fig. 2.1) (Kyuma 2005). It is only when these reach a sufficient magnitude that we can

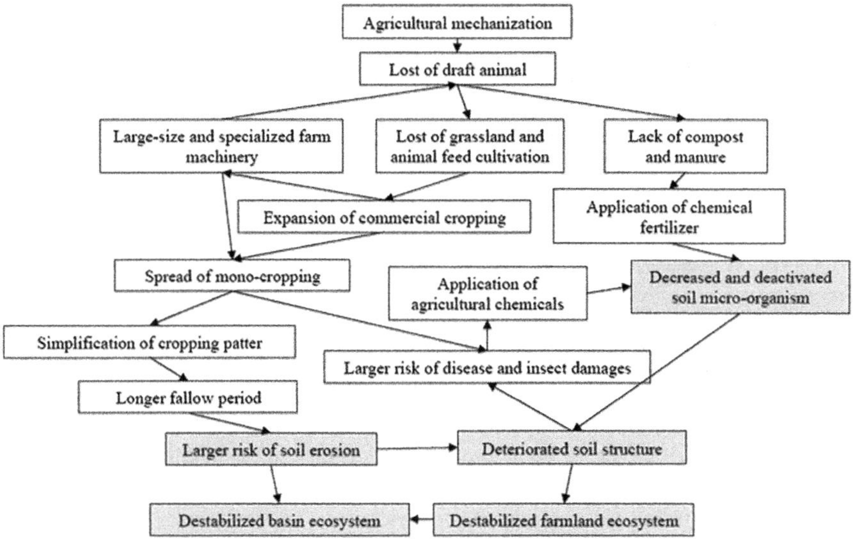

Fig. 2.1 A cascade of changes in the local geosphere and biosphere induced by input-intensive agriculture. (*Remarks* Adapted from Kyuma 2005)

perceive them as feedback in terms of degrading the productive capacity of the environment and reducing its productivity.

The more extensive the practice of agricultural development, the more intensive the effects. In Pakistan, immediately following its independence, a vast area of desert land was converted to agricultural land through large-scale irrigation development based on the transboundary water transfer of the Indus River. The command area of the Indus River irrigation system spans approximately 16 million ha. Today, due to rising groundwater levels and excessive evaporation, approximately 2 million ha of farmland now suffer from the effects of serious salinization, with the area increasing by 30–40 thousand hectares a year (Aslam and Prathapar 2006). In Indonesia, in the latter half of the 1990s, in order to promote transmigration from an overpopulated Java Island, an attempt was made to reclaim 1.4 million ha of tropical peat swamp land on Kalimantan Island, an effort known as the Mega Rice Project (MRP). The only result, however, has been the loss of vast quantities of peat, exposing underlying acidic soil leading to the production of a vast expanse of barren land (Furukawa et al. 2004).

Scholars have noted that agriculture is humanity's "original sin" (Tudge 1999). Historically, agriculture not only created a discrepancy in wealth but also produced conditions that led to the subjugation of other groups, leading to the creation of social contexts, where women's status was lower than that of men leading to the development of class societies. In addition, it also served as the vehicle for extensive intervention in both the geosphere and biosphere through land reclamation projects, water management, and the extermination of wild animal populations.

Here, we do not examine the question of agricultural practices itself. Human societies have tested and elaborated a wide range of farming systems and repeatedly experienced success and failures, some with fatal outcomes for local societies (Diamond 1997). This trial and error process improved farming systems to respond and better adapt to local environmental conditions. Such efforts and innovations have supported the continued increase in human populations and improved living standards over the course of 10,000 years of human history (Bellwood 2005).

Yet, during the last several decades, agricultural development practices have had a profound impact on the geosphere and biosphere. They have impacted at a hitherto unknown level with the aim of controlling the production environment. This has resulted in huge impacts at both a local and global level in the present and will transform the future geosphere and biosphere (McNeill 2000). Nevertheless, these practices have not been tested over a sufficient period of time. More importantly, as will be argued in this chapter, insufficient value has been placed on the importance of the trial process. When the cascading effects in the geosphere and biosphere are perceived as unfavorable conditions, they tend to be treated as the symptoms of unfavorable impacts, ones which can be ameliorated through additional investment and technical adjustments, rather than exploring the black box of the geosphere and biosphere, verifying underlying causes, and reorganizing agricultural technology. This tendency has more serious implications in the tropics, where the geosphere and biosphere are much more diverse and complicated than in the temperate zone (Kanzaki and Yamada 2010; Yanagisawa et al. 2012).

In this chapter, we focus on different types of agricultural development in the tropics and discuss the appropriateness of their timescales from a sustainability perspective. Agriculture practiced in the tropics is extremely diverse and intensive modern technology-driven commercialized farming coexists with traditional subsistence farming practices dependent on natural cycles. Neither should be labeled unconditionally as being superior or inferior, as each has emerged within different historical continuums under different economic circumstances and exist in different environmental contexts. Two case studies in mainland Southeast Asia, the development of rainfed paddy cultivation in the Khorat plateau, Thailand and the expansion of commercial crop plantations near the border between Laos and China, are used to illustrate how agricultural practices in mainland Southeast Asia have changed rapidly over the past decade and come to play a role in the rise of the Greater Mekong Subregion (GMS) and the rapidly evolving economic development conditions of the region. While these changes may be socially and politically inevitable (and unfortunately, irreversible), this chapter proposes an integrated framework based on micro-development mechanisms in realizing sustainable agricultural development in the tropics.

Micro-Development of Agriculture

Farming is not an activity that involves simple repetition but constantly develops and evolves in response to a production environment that consists of a constantly changing local geosphere and biosphere. Year-to-year fluctuations and changing seasonality of the climate deeply affect the progress of farming practices, crop growth, and various kinds of stresses on crop, as well as the occurrence of natural disasters. As such, the social, economic and household conditions of farmers are continually undergoing transformations. In the case of the cultivation of commercial crops, farmers, keeping an eye on market prices, think about when and what to plant. As farmers age, they tend to avoid strenuous work; they may sell water buffalos to buy power tillers according to changes in the availability of family labor. Farming is, in reality, a continuous adaptation and adjustment toward the local conditions of the geosphere and biosphere as well as societal conditions. This is one of the clear differences between agriculture and industrial manufacturing.

Such farm- and household-level skills and small ingenuities are not generally considered to form part of agricultural development and are not listed on agricultural research agendas. However, farming practices represent an important intersection of human society, the geosphere, and biosphere. Farmers, while paying heed to the needs and constraints of their families, villages, markets, and nations, do not simply attempt to gain the highest production and profit, but rather pay close attention to the signs that the geosphere and biosphere disclose, attempt to decipher the underlying causes and consequences of their movements, and strive to find a harmonized option that minimizes risks, maximize immediate outcomes, and secure life and livelihood. This process of examination is not systematized and merely represents an accumulation of suboptimal choices. What is prioritized is not the achievement of predetermined goals, but rather an adaptation to a wide range of changing conditions. We define these mechanisms of changes as "micro-development." Development here means to reconstruct the knowledge and technology for practice and consequently of human society while taking the geosphere and biosphere into consideration as the foundation for sustainable life (in the broad sense) and livelihood.[1] We believe that farm- and household-level mechanisms will contribute to this end.

[1]Some readers may feel it is inappropriate to use the term "development." However, we are attempting to prioritize sustainable life and livelihood rather than locating higher production and productivity as the central issue of development. This is why we chose to use the term "development" in reference to life, i.e., its diversity in the biosphere upon which human society depends on, and how that supports human livelihood.

Development of RainFed Paddy Cultivation in the Khorat Plateau, Thailand

High population density in Monsoon Asia has been sustained by paddy cultivation whose crucial lifeline is water. Land with alluvial soils is repeatedly replenished by abundant and highly seasonal rains and frequent flooding in core regions is crucial for paddy cultivation. In contrast, the productivity of paddy in semiarid regions is substantially lower as they experience relatively low precipitation of less than 1000 mm per year, are poor in alluvial soil, and consist largely of erosion surfaces (Fig. 2.2) (Kono 2009). They represent marginal lands for paddy cultivation; water is a limiting factor; and the ability (or inability) to secure water is the key determinant not only for the success (or failure) of paddy cultivation but also for the paddy farmers' very survival. If droughts continue for several years, rice stocks are exhausted, and people are forced to venture into forests to find food or to visit nearby villages to beg for food. The methods to secure water vary from location to location yet, in the Khorat plateau, farmers primarily harvest water and accumulate rainfall in fields. This is a substantially different approach when compared to their counterparts elsewhere, as in the Deccan plateau, India, where people construct tanks (Sato and Ramasamy 2011).

Until the 1980s, paddy cultivation of the Khorat plateau was accomplished through an elaborate combination of on-farm water management and rice cultivars,

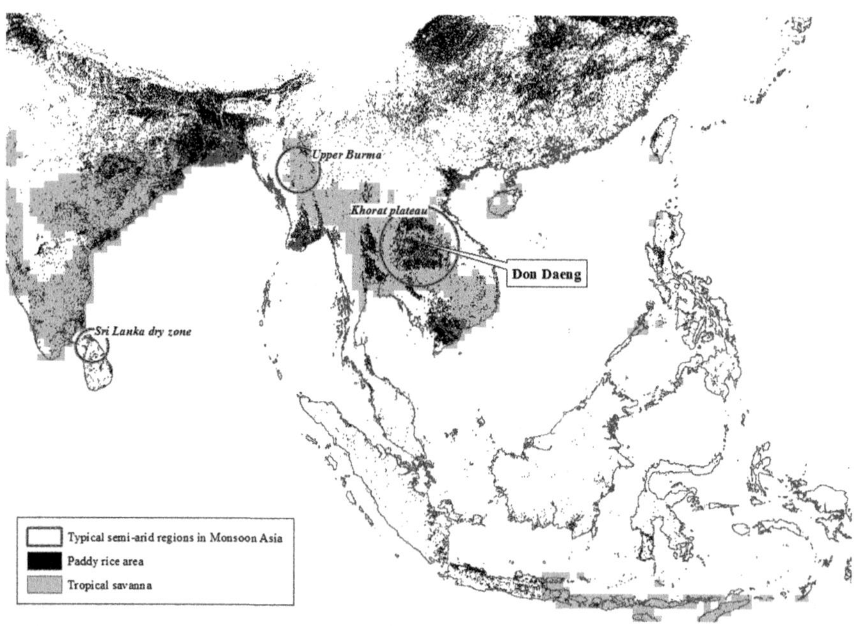

Fig. 2.2 Distribution of paddy cultivation in Southeast and South Asia (*Remarks* Adapted from Nelson and Gumma 2015 and Kottek et al. 2006)

forming a technological system adapted to the region's water environment (Fig. 2.2) (Kono et al. 2010). This rainfed paddy cultivation was sustained within the context of an overall livelihood system combining, among others, dry land farming, vegetable gardening, animal husbandry, a wide variety of day labor, and work away from home during the dry season. Thai society has experienced rapid economic growth on the order of 10% per year since the mid-1980s. The capital city, Bangkok, developed into a megacity, resulting in the emergence of a middle class engaged in the manufacturing and service sectors. The effects of this economic growth and the rise of GMS also extended to rural areas in Thailand and to villages on the Khorat plateau. The practice of working outside villages increased, particularly among young people who were being employed as factory workers, returning home only a few times a year during holidays, and leaving behind only elderly individuals and children in their villages of origin (Funahashi 1996). The Thai government accelerated implementing measures to upgrade the living standard and to enhance social welfare of farmers through improving infrastructure as well as educational and health care systems in the mid-1990s. How, then, has rainfed paddy cultivation responded to these socioeconomic changes?

In the early 1980s, researchers engaged in research on rainfed paddy cultivation on the Khorat plateau (Kono et al. 2009), clarified that the farming system of rainfed paddy in place was already optimally adapted to the existing water environment (Fukui 1993). Research predicted that the increase in off-farm job opportunities resulting from national-level economic growth would lead to the abandonment of rainfed paddy cultivation (Kono et al. 2009). Yet, changes over the last 20 years or so have completely upended this prediction. Undoubtedly, off-farm income source has come to account for the majority of farmers' labor and remittances from family members working away from the village have come to play an important role in the household economy (Funahashi 2006). However, these changes did not result in the abandonment of paddy cultivation. In fact, paddy cultivation has been strengthened and undergone a transformation to become the basis of household solidarity and security. During the Asian monetary crisis of 1998, the majority of migrant workers returned to their villages of origin which provided them with a safety net (Krishnamurty 2009). Rice yields, which were only 1–2 t/ha in the early 1980s, increased to approximately 3 t/ha in the early 2000s (Fig. 2.3). Droughts, which were experienced every other year in the 1980s, became less frequent in the 1990s, occurring once every five years (Table 2.1) (Kono et al. 2009). Grandstaff et al. (2008) refer to this development as the "rainfed revolution."

Through what mechanism was this "revolution" realized? It was a consequence of numerous small changes of farm- and household-level skills and tunings: changes that were not prepackaged, but rather sporadic. Crucially, they were changes that were not consistently designed, but rather suboptimal to site- and time-specific conditions. This is not a result of the collective actions of farmers, though they do observe and learn from each other. These changes were observed over the last 20 years at a farm village near Khon Kaen, a center of Northeast Thailand, which began to experience the impacts of rapid economic growth in the

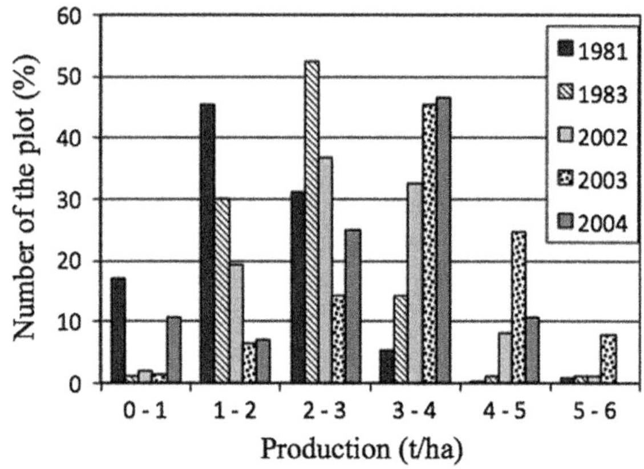

Fig. 2.3 Changes in plot-level rice yield distribution at the village in Khorat plateau. (Revised from Funahashi 2006)

Table 2.1 Frequency of occurrence of drought and flood damage at a village in northeastern Thailand between the 1980s and the 2000. (Kono et al. 2009)

Kind of damage	Damaged area	Frequency	
		1978–1990 (13 years)	1991–2002 (12 years)
Drought	Almost whole area	3	0
	About half area	1	0
	Partial	4	2
Food	Almost whole area	2	1
	About half area	0	1
	Partial	0	3

1990s (Konchan and Kono 1996). As a result, research confirmed that rainfed paddy cultivation underwent a transformation.

The first change was land consolidation. In the past, lowland fields were delineated according to the original terrain in order to minimize civil works for reclamation. Each field was small and terraced. Some of these fields were merged to form fewer, but larger fields through the use of bulldozers. The average number of fields owned by a household decreased from 8.4 (in 1981) to 2.9 (in 2005) (Watanabe et al. 2008). In effect, this contributed to effective farm-level water management. More importantly, the efficiency of farm work was improved through mechanization, where power tillers and harvesters were widely adopted and contributed to labor savings.

In addition to the above changes, supplementary irrigation was also introduced. The government installed a pump station on the Chi River, a tributary of the Mekong River flowing north of the village, and constructed an irrigation canal to the village's lowland fields. Irrigation water is thus supplied according to the villagers' requests, but they have to pay for the costs of pump operation (Hoshikawa and Watanabe 2006).

Lowland fields located alongside the irrigation canal have come to use irrigation water, however, even with the most generous estimates, these account for only 40% of the village's total lowland field area (Fig. 2.4). Fields are not regularly irrigated, as would be common practice for a typical irrigated paddy. Villagers request this additional irrigation only when droughts continue and there is a fear that planting will be delayed, or when the plants are beginning to show signs of drought stress. The remaining lowland fields are also irrigated by pumping harvested water at lower fields and/or swamps. Neither of these two groups of lowland paddy is entirely irrigated nor entirely rainfed. Water, then, is managed while keeping an eye on the progress of farming practices, on the growth of paddy, on the water supply, and on the weather.

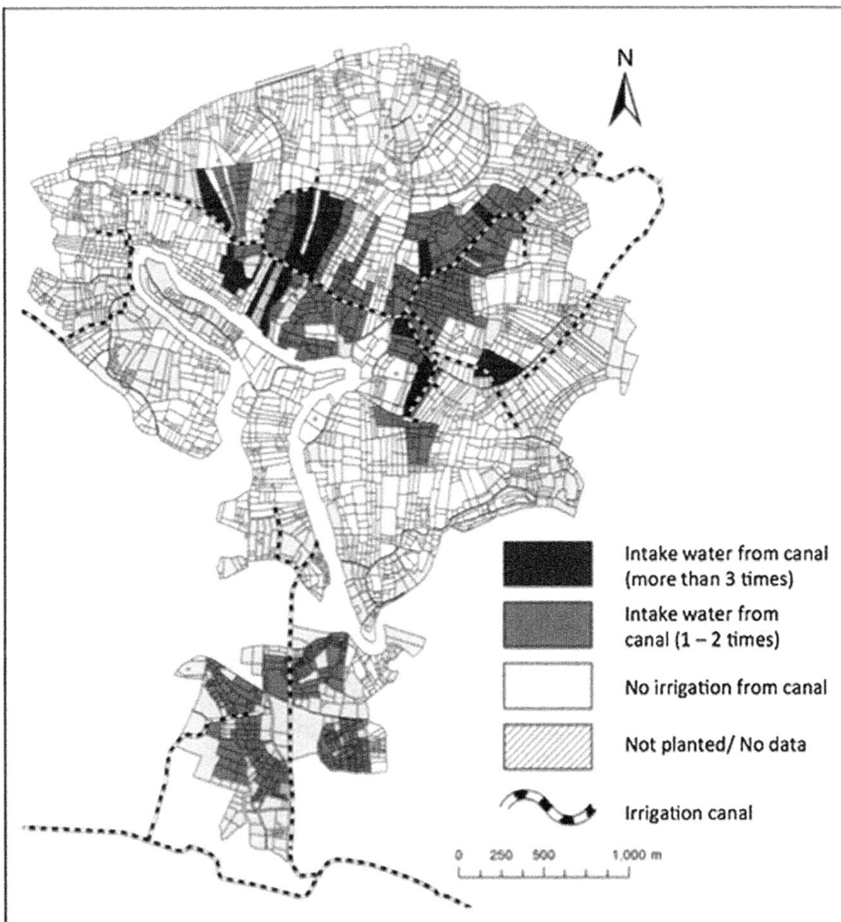

Fig. 2.4 Irrigation water used for paddy cultivation at the village in the northeastern Thailand. (Revised from Kono et al. 2009)

The varieties of rice cultivars have also changed. Approximately 30 varieties, differing in heading period, were cultivated in the village's 360 ha of lowland fields, in the early 1980s. By the 1990s, these were replaced by two medium-heading varieties, a high-yielding glutinous rice (RD6) and a superior non-glutinous rice (KDML105), which occupied nearly 90% of the total lowland field (Miyagawa 1995). These two varieties were primarily selected for their good taste. The coordination of heading period contributed to save on harvesting labor. Furthermore, planting methods also changed, from the predominant transplanting of seedlings to a combination of transplanting and direct seeding. The reason for this change is, again, labor saving (Konchan and Kono 1996). Direct seeding into dry fields results in the greater emergence of weeds. In order to deal with this problem, farmers have attempted to rotate transplanting or resorted to using of herbicides. Chemical fertilizers have also come to be used. While these contributed to uplifting yields to some degree, their primary effect was to improve the growth of poorly growing plants. As such, this does not lead to the establishment of a standardized application practice, but rather the use of small quantities of fertilizer as and when needed.

The above are examples of small changes that have been tried and tested since the 1990s. These small changes are not expected to drive revolutionary increase in productivity. The approach that was observed is that farmers pay continuous attention to the particular characteristics of their fields, carefully observe responses of the land and crop to their trials, and fine-tune their trials. These efforts result in a much wider disparity in yields among plots, from 0 to 3 t/ha with the average of 1 t/ha in the early 1980s to 1–6 t/ha with average of 3 t/ha in the early 2000s (Fig. 2.3). This data evidently suggests that the mechanism of paddy cultivation development of this village is neither a unification nor standardization of their farming system, but an accumulation of site- and time-specific micro-developments. Higher productivity is not always a central issue but one of the concerns of micro-development.

From Subsistence Paddy to Commercial Crop Plantations

The continental mountainous region of Southeast Asia comprises of both mountainous slopes and intermountain basins. Swidden agriculture is practiced on slope land and lowland paddy farming in the intermountain basins. While political affairs in many parts of Mainland Southeast Asia have resulted in the relocation of inhabitants of mountainous regions and forced them to experience the repeated disruption and reconstruction of social order, agricultural systems have remained essentially unchanged for several hundred years (Daniels 2008). Until recently, this region consisted of isolated "islands" with a long history of caravans weaving their way, to and from different mountainous areas (Kato 1998; Daniels 2008). Goods transported in this manner were, in addition to tea, salt, and opium and high-value forest products such as benzoin, cardamom, and animal pelts (Akimichi 2009).

In recent years, however, this region has gained prominence as a corridor connecting China to Southeast Asia. A road network crossing the region from north to south and from east to west has been constructed under the regional framework of the GMS (Krongkaew 2004). One of these thoroughfares is the highway connecting Kunming city, the capital of Yunnan Province, to Bangkok, the capital of Thailand. Construction of the road began in Kunming and arrived in the southern tip of Xishuangbanna in 2007, and is presently connected to Bangkok via Laos. This corridor allows for the flow of people through and across borders, but its primary purpose is for the transportation of goods, especially commercial crops from Mainland Southeast Asia, to satisfy market demand in China. With this in mind, we look, in detail, at two different villages located near the border between Laos and China (Fig. 2.5).

One village on the Lao side is peacefully nestled in an intermountain basin and has its livelihood basis of paddy cultivation with weir irrigation (Tomita et al. 2008). The road connecting this village with the Chinese border, situated 15 km far away, was repaired in 2001, which resulted in an increase in traffic to and from the village. Numerous Chinese traders started to visit and encouraged villagers to produce a wide range of commercial crops for Chinese market (Fig. 2.6). The negotiating process between Chinese traders and Lao farmers upon the introduction of watermelon, a commercial crop, proceeded as follows.

The village was visited by four Chinese brokers in 2006 who gave a presentation in the village's assembly hall. The village headman at the time decided that

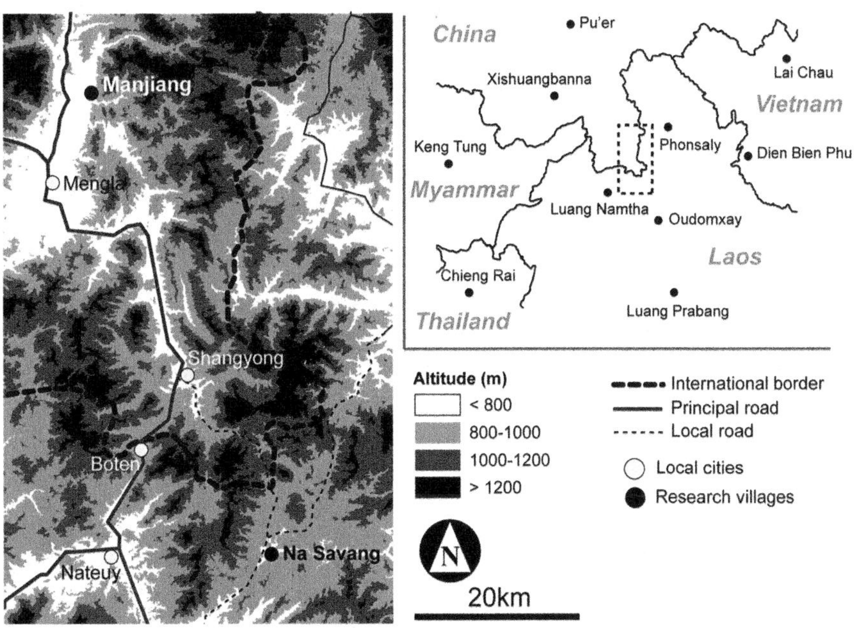

Fig. 2.5 Map of two villages (Laos-China border)

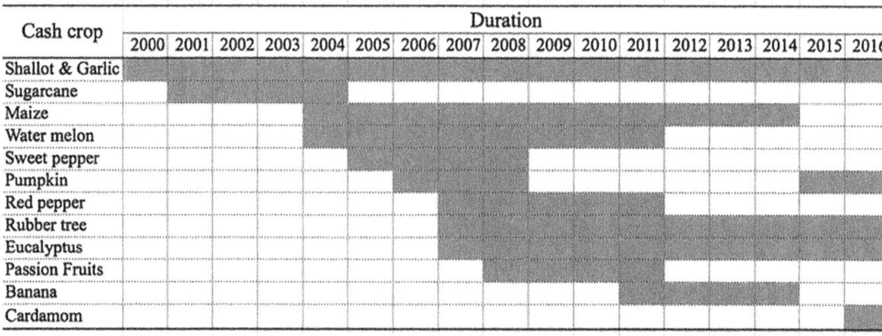

Fig. 2.6 Introduction process of the commercial cropping for Chinese market at the northwestern Lao village

contracts should be made between the brokers and individual households, not the village as a whole, and left the decision of which broker to take out a contract with each household. Approximately 20 village leaders entered into such contracts and attempted to grow watermelons as a second crop after paddy. One of the brokers dispatched an instructor to teach the villagers watermelon cultivation techniques; the instructor was stationed in the village, providing villagers ample opportunity to learn relevant cultivation techniques. Brokers came back to the village around the time of harvest, however, three either returned to China without paying or only paid half of the contracted compensation. The only broker who faithfully fulfilled the contracts was the one who had dispatched the technical consultant. This broker continued to bring contracts for watermelon until 2009 but did not appear in 2010. This is believed to have been the result of the tighter custom management of cross-border trade. In response, villagers attempted to sell their products in Laos.

The introduction of other commercial crops followed a similar process. There are cases of sugarcane, green pepper, and pumpkin whose cultivation was abandoned after two or three years and also cases of onion and garlic whose cultivation have started to take root. The area of land given over to planting any of these crops is small and subsistence wet rice farming is still at the very center of villager's livelihood, although most households are engaged in commercial crop cultivation. Consequently, the impacts of introducing commercial cropping upon the village economy as well as landscape are not outstanding.

What the above makes clear is that the process of negotiation between Laotian farmers and Chinese brokers was not conducted on the basis of clear and transparent rules. As a result, both parties attempted to find a common ground that was to their mutual benefit, even while harboring suspicion toward each other. The creation of any rules governing trade will take time and, above all, will require that Lao farmers learn about the Chinese market and that the Chinese brokers learn about the Lao farmers and their production environment. Decision-making is not collective and is not supported by any public institutions; rather, both Laotian farmers and Chinese brokers need to rely on their own acumen at each step of the process, while keeping an eye on cross-border trade regulations. Such a process is generally inefficient. However, it may well be the case that just such a process can potentially allow both Chinese brokers and Lao farmers to spend enough time to identify suitable crops and cropping technology specific to the local environment and market conditions and to foster effective knowledge in forging a sustainable livelihood basis.

Parallel to the transformations above, a different type of agricultural development took place in the Chinese side of the border. In the early 1990s, the basins of Xishuangbanna, Yunnan Province, were quiet and peaceful as it was on the Lao side. Plentiful irrigation water was diverted to fill all paddy fields and produced sufficient amounts of rice. Village settlements along these basins consisted of sturdy houses topped by tiled roofs with long-extending eaves. Located in a remote region of China, these villages were not necessarily blessed in the economic sense, but they represented a robust sustainable life in their own right.

The construction of the corridor highway triggered a drastic change in agriculture. The highway connected quiet Xishuangbanna villages to Han society and China's domestic market and promoted a wide range introduction of commercial crop plantation. At a Dai village located near the Laos border, banana cultivation was introduced in 2008 (Fig. 2.5). By 2010, 63 ha of the 65 ha of paddy fields had been converted for banana cultivation and 97 of 100 households were engaged in banana production (Fig. 2.7) (Zhang et al. 2014). In other words, in as little as three years, almost all the farm households had switched livelihood from subsistence paddy production-based to commercial cropping-based. Why did this shift of livelihood occur so rapidly? First, the profitability of banana cultivation is overwhelming higher than rice cultivation. Bananas are piled into trucks and distributed to markets throughout the country. While there is a tremendous demand for bananas in China, land suitable for its cultivation in the country is limited. Second, Han people including technicians, fertilizer and agricultural chemical salesmen, and distributors rushed into Xishuangbanna and brought with them a pre-assembled system for banana cultivation and distribution as well as production support. In

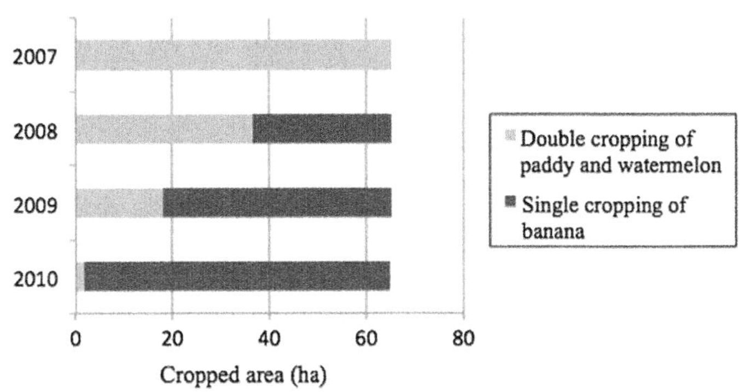

Fig. 2.7 Changes in cropping pattern of lowland field at a village in Xishuangbanna. (*Source* Zhang et al. 2014, Fig. 2.2)

some cases, banana cultivation is carried out on land leased from local farmers by Han banana producers coming from Sichuan and Guangdong Provinces. Escaping the highly competitive business environment of central China, these migrants are seeking business opportunities in a relatively less-competitive frontier region. Taking the above into consideration, it may seem to be rational to quickly switch livelihood from subsistence to market-dependent ones.

The contrast of two cases in both sides of national border suggests that different types of agricultural development can coexist even under a similar ecological setting. What happened on the Chinese side, led to a greater amount of commercial production, on one hand providing more income for farmers and promoted a quick shift in farmers' livelihood from subsistence to market-dependency. On the other, changes on the Lao side included a repeated trial and error approach: typical characteristics of micro-development with an outcome that is less explicit and visible. The primary cause behind this difference is the existence of powerful external drivers of development, prepackaged advanced technologies of production, processing and distribution supported by large markets, and solid institutions. The external driver effectively guides farmers to adopt new types of farming and to yield high economic returns on the Chinese side but takes away the opportunity for farmers to sufficiently consider and evaluate impacts on the geosphere and biosphere. It is still not clear how the massive use of chemical fertilizers and pesticides in the cultivation of bananas is affecting the regional agroecology and residential environments of villages in Xishuangbanna, which should be carefully monitored and further investigated.

Agricultural Development Timescales

Human societies share a long history of learning about the structure and movement of the geosphere and biosphere. Interactions and observations within an empirical academic framework have produced a wide range of compartmentalized disciplinary approaches ranging from geology and climatology to biology and ecology. These disciplines have mainly been elaborated by temperate world-based researchers, developed in a specific continuum, and refer more specifically to records and observations there. In a sense, theories have consequently been more suited to understandings and conceptualizations of the geosphere and biosphere in the temperate world. The tropics still remains a frontier for many disciplines and available knowledge on the geosphere and biosphere has yet to truly produce a comprehensive approach attuned to the complexity of the environment and human-nature interactions.

The geosphere and biosphere in the tropics have a more complex and spatially diverse structure than those in the temperate zone. Tropical agriculture has to confront the geosphere and biosphere and utilize agricultural resources in the face of richer diversity and higher uncertainty. Micro-development is one form of practical and effective adaptation to environmental conditions. The standardized approach of using packaged technology has been more or less successful in the temperate world with a moderate geosphere and biosphere, but it is doubtful whether this technological approach can be fully effective in the tropics. A deeper and wider intervention into the geosphere and biosphere is required when we apply this to the tropics. The cascade of effects and impacts on regional human societies are difficult to predict, partly because of robust understandings on and partly because of the complex, diverse, and vigorous characteristics of the tropical geosphere and biosphere. Crucially, irrespective of the impacts, due to the importance of the tropics for the earth system these have the potential to spread beyond the current tropics to influence the entire planet and the future foundations for intra-species survivability.

What this discussion suggests is that agricultural development in the tropics should suppose diverse optimums in different localities and development processes. In other words, the understanding of the local geosphere and biosphere should be the starting point of any development process as we observed in the process of micro-development. As such, we need two kinds of engines, one that cares for the geosphere and biosphere and another that proactively adapts to them, and enhance the development mechanisms through two complementary operating engines, as observed through micro-development (Fig. 2.8).

What, then, does it mean *to care* for the geosphere and the biosphere? In general terms "care" means to "take care of," "look after", or to "provide nursing assistance." The precondition for care is that there is a relationship between the party providing the care and the party receiving the care predicated on a mutual recognition of and respect for the other's needs. In the broader sense, however, care means to consider and to proactively work toward improving the well-being of the other party. The "other party" should not only be limited to one particular species, homo sapiens, but also flora, fauna, or the physical environment (Fisher and Tronto 1990; Hayami et al. 2012). Caring for the geosphere and biosphere means recognizing them as coexisting domains that are necessary for our survival and striving to comprehend their laws, structure, and movements.

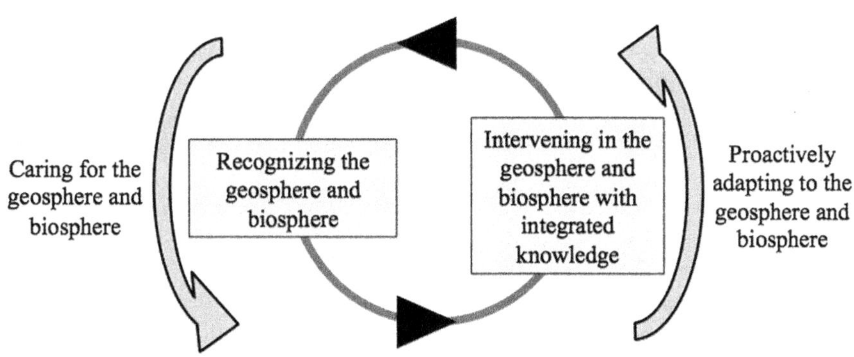

Fig. 2.8 Mechanism for sustainable development of tropical agriculture

Until relatively recently, a perception has existed that the geosphere and biosphere are something beyond the scope of the human intellect. They are a source of natural disasters and venerated as a fundamental requirement of our existence derived from our experiential knowledge. On the other hand, we have seen inordinate efforts to monitor the movements of geosphere and biosphere through modern technology and to create greater scientific knowledge of their natural laws. To care for the geosphere and the biosphere means to take approaches represented by these two extremes and, by combining them in a complementary manner in order to strive for a more integrated knowledge of both.

What does it mean to proactively adapt to the geosphere and biosphere? Adaptation is an essential survival strategy limited to human society, but of many forms of life in general. Human society, however, has grown to not only adapt to but also transform the geosphere and biosphere. To proactively adapt means, therefore, to create technologies and institutions that interact with the geosphere and biosphere based on integrated knowledge. Stakeholders of agricultural practices are not only farmers, but also researchers, business sectors, and policy makers working on agriculture. The two empirical examples in this chapter show how we must root our understandings in how the geosphere and biosphere function within the very specific characteristics of the tropics. To apply a temperate-climate "mindset"—or logic—upon the tropics will lead to an abrupt transformation in the foundations for survival for not just human societies but also other species, especially those that provide crucial ecosystem services for crops, forests, and the ecological climates that form Southeast Asia.

Conclusion

Agricultural development based on standardized and packaged technology, which was the major driver that led to an increase in agricultural production in the latter half of the twentieth century, might be an emergency measure to respond to an unprecedented explosion of population growth and people's demands on food and materials. Over the past 70 years, there was no other choice but to adopt this development approach with a short-term perspective and to prioritize short-term production. With little margin to operate an "engine of care," we have been forced to operate an engine of intervention without a clear idea of consequences. Unfortunately, the logic of such unbalanced development still holds sway in the tropics today.

This leads us to a question. How do we synthesize scientific and experiential knowledge? This is by no means a new question for human societies. Under the conditions requiring the prioritization of the standardized approach to development, agricultural researchers had to choose a way to deepen finely divided disciplinary knowledge. They were not allowed to step outside disciplines to integrate the two engines, elucidated above, into one mechanism. Farmers have paid attention to care for and to proactively adapt to their local geosphere and biosphere through micro-development and have produced experiential knowledge. Yet, their activities were often sidelined by orthodox research, recognized as outdated practices, and not given the opportunity to contribute to the reconstruction of agricultural technology and institutions suited to the tropics. By relativizing experiences of agricultural development under the explosive expansion of human societies and reconsidering a suitable timescale of perspectives for taking the geosphere and biosphere into consideration, we open up the possibility of removing the barriers dividing scientific and experiential knowledge. To accumulate countless century-old small efforts and fairly evaluate them will be crucial for realizing sustainable agricultural development in the tropics for the twenty-first century.

References

Akimichi, T. (Ed.). (2009). *An illustrated eco-history of the Mekong river basin*. Bangkok: White Lotus.

Aslam, A., & Prathapar, S. A. (2006). *Strategies to mitigate secondary salinization in the Indus basin of Pakistan: A selective review*. Colombo: International Water Management Institute.

Bellwood, P. (2005). *First farmers: The origins of agricultural society*. Oxford: Blackwell Wiley.

Daniels, C. (Ed.). (2008). *Ronshu monsun Ajia no seitaishi Dai 2 maki Chi-iki no seitaishi* [Eco-history of monsoon Asia Vol. 2. The eco-history of local Society]. Tokyo: Kobundo (in Japanese).

Diamond, J. (1997). *Guns, germs, and steel: The fates of human societies*. New York: Norton.

Fisher, B., & Tronto, J. (1990). Toward a feminist theory of caring. In F. Abel & M. Nelson (Eds.), *Circles of care* (pp. 35–62). New York: State University of New York.

Food and Agriculture Organization of the United Nations (FAO). Statistical database (FAOSTAT). Retrieved August 4, 2016, from http://faostat3.fao.org/home/E.

Fröhlich, H. L., Schreinemachers, P., Stahr, K., Clemens, G. (Eds.). (2013). *Sustainable land use and rural development in Southeast Asia: Innovations and policies for mountainous Areas*. Heidelberg: Springer.

Fukui, H. (1993). *Food and population in a northeast Thai village*. Honolulu: University of Hawai'i Press.

Funahashi, K. (1996). Farming by the older generation: The exodus of young labor in Yasothon Province, Thailand. *Southeast Asian Studies, 33*(4), 625–639.

Funahashi, K. (Ed.). (2006). *Don Daeng Mura Saisaiho: Hokuto Tai tensuimitsu noson ni 40 nen no idokenkyu*. [Don Daeng village Re-revisited: Four decades of dynamism studies of a rainfed paddy growing village in northeast Thailand]. Kyoto: Ryukoku University (in Japanese).

Furukawa, H., Nishibuchi, M., Kono, Y., & Kaida, Y. (2004). *Ecological destruction, health and development*. Kyoto: Kyoto University Press.

Grandstaff, T. B., Grandsutaff, S., Limpiuntana, V., & Suphanchaimat, N. (2008). Rainfed revolution in Northeast Thailand. *Southeast Asian Studies, 46*(3), 289–376.

Hayami, Y., & Kikuchi, M. (2000). *A rice village saga: Three decades of green revolution in the Philippines*. London: Macmillan.

Hayami, Y., Nishi, M., & Kimura, S. (2012). *Ningenken no saikochiku: Nettashakai no senzairyoku [Reconstituting the human domain: The potentialities of tropical societies]* Kyoto: Kyoto University Press (in Japanese).

Hoshikawa, K., & Watanabe, K. (2006). Changes in agricultural infrastructures in Don Daeng Village. In K. Funahashi (Ed.), *Don Daeng Mura Saisaiho: Hokuto Tai tensuimitsu noson ni 40 nen no idokenkyu*. [Don Daeng village Re-revisited: Four decades' of dynamism studies of a rainfed paddy growing village in northeast Thailand]. Kyoto: Ryukoku University (in Japanese).

Kanzaki, M., & Yamada, A. (2010). *Seitai kiban toshite no seibutsu tayosei.* [Biodiversity as the sustainable humanosphere]. In K. Sugihara, S. Kawai, Y. Kono, & A. Tanabe (Eds.), *Chikyuken, seimeiken, ningenken: Jizokutekina seizon kiban wo motomete. [Geosphere, biosphere and human society: Searching for sustainable humanosphere]* (pp. 153–184). Kyoto: Kyoto University Press (in Japanese).

Kato, K. (1998). *Shisonbanna no koekiro.* [Trade routes of Xishuangbanna]. In T. Shintani (Ed.), *Ogon no shikaku chitaki: Shan bunkaken no rekishi, gengo, minzoku* [The golden square: History, language and ethnicity of Shan culture zone]. Tokyo: Keiyu-sha (in Japanese).

Konchan, S., & Kono Y. (1996). Spread of direct seeded lowland rice in northeast Thailand: Farmers' adaptation to economic growth. *Southeast Asian Studies, 33*(4), 523–546.

Kono, Y. (2009). *Hankan sochiiki no inasaku* [Rice cultivation in semiarid regions] In S. Haruyama, M. Fujimaki, & H. Noma (Eds.), *Asakura sekai chiri koza 3 Tonan Ajia* [Asakura World Geography Vol. 3 Southeast Asia] (pp. 167–179). Tokyo: Asakura Shoten (in Japanese).

Kono, Y., Miyagawa, S., & Watanabe, K. (2009). Interpreting social changes through a village level paddy productivity map. In T. Mizushima & M. Shibayama (Eds.), *GIS for Area Studies* (pp. 81–93). Tokyo: Kokin Shoin.

Kono, Y., Song, S., & Hoshikawa, K. (2010). *Mizu no riyo kara mita nettai shakai no tayosei* [The diversity in water use of tropical societies]. In K. Sugihara, S. Kawai, Y. Kono, & A. Tanabe (Eds.), *Chikyuken, seimeiken, ningenken: Jizokutekina seizon kiban wo Motomete* [Geosphere, biosphere and human society: Searching for sustainable humanosphere] (pp. 185–209). Kyoto: Kyoto University Press (in Japanese).

Kottek, M., Grieser, J., Beck, C., Rudolf, B., & Rubel, F. (2006). World map of the Köppen-geiger climate classification updated. *Meteorol. Z., 15,* 259–263. https://doi.org/10.1127/0941-/2006/0130.

Kozan, O. (2010). *Chikyuken no kudoryoku toshite no nettai.* [Tropics as the Driving Force of Geosphere] In K. Sugihara, S. Kawai, Y. Kono, & A. Tanabe (Eds.), *Chikyuken, seimeiken, ningenken: Jizokutekina seizon kiban wo motomete* [Geosphere, biosphere and human society: Searching for sustainable humanosphere] (pp. 129–152). Kyoto: Kyoto University Press (in Japanese).

Krishnamurty, J. (2009). Learning from the 1997–1998 Asian financial crises: The ILO experience in Thailand and Indonesia, employment report No. 3. ILO.

Krongkaew, M. (2004). The development of the Greater Mekong Subregion (GMS): Real promise or false hope? *Journal of Asian Economics, 15*(5), 977–998.

Kyuma, K. (2005). *Tsuchi to ha nan darou ka* [What is soil?]. Kyoto: Kyoto University.

McNeill, J. R. (2000). *Something new under the sun: An environmental history of the twentieth-century world.* New York: Norton.

Miyagawa, S. (1995). Expansion of an improved variety into rain-fed rice cultivation in northeast Thailand. *Southeast Asian Studies, 33*(2), 187–203.

Nelson, A., & Gumma, M. K. (2015). A map of lowland rice extent in the major rice growing countries of Asia. http://irri.org/our-work/research/policy-and-markets/mapping.

Sato, T., & Ramasamy P. (2011). The effects of expansion of private wells on rural livelihood in tank intensive watersheds: A case study in upper Gundar River Basin, Tamil Nadu. *Southeast Asian Studies, 49*(1), 124–150.

Tomita, S., Kono, Y., Kotegawa, T., & Chowdary, B. M. (2008). *Tonan ajia tairiku sanchibu no tochi riyou no gijutsu to chitsujo no keisei* [Technology and regime formation of land use in the mountainous region of Southeast Asia]. In C. Daniels (Ed.), *Monsun Ajia no seitai shi, Chiiki no seitai shi* [Eco-history of Monsoon Asia Vol. 2 Eco-history of local society]. Tokyo: Kobundo (in Japanese).

Tudge, C. (1999). *Neanderthals, bandits and farmers: How agriculture really began.* New Haven: Yale University Press.

United Nations Population Division. Retrieved July 27, 2016, from http://www.un.org/en/development/desa/population/.

Watanabe, K., Hoshikawa, K., & Miyagawa, S. (2008). *Taikoku tohokubu Don Daen mura in Okeru tensuiden no kugakaihen to sono suito seisan he no eikyo* [Farm layout improvement and its impact on paddy productivity at Don Daeng village]. *Nogyo nozon kogakukai ronbunshu* [Northeast Thailand. Journal of Irrigation Engineering and Rural Planning], *76*(1), 45–52.

Yanagisawa, M., Kono, Y., Kozan, O., & Kanzaki, M. (2012). *Chikyuken, seimeiken no Senzairyoku: Nettai chiikishakai no seizonkiban* [The potentiality of geosphere and biosphere: Exploring the tropical humanosphere]. Kyoto: Kyoto University Press (in Japanese).

Zhang, L., Kono, Y., & Kobayashi, S. (2014). The process of expansion in commercial banana cropping in tropical China: A case study at a Dai village, Mengla County. *Agricultural Systems, 124,* 32–38.

Chapter 3
Fueling Transformation in the Mekong: Thailand's Trade in Agro-Energy

Sabrina Shaw, Samai Jai-In and Muanpong Juntopas

Abstract The Mekong region stands at an important development juncture, where the dynamics of sustainability are being defined through an unprecedented level of cooperation on agriculture, energy, and trade in combination with action to address climate change. This chapter examines the policy space to develop sustainable agriculture while fueling transformation in the Mekong region based on research from the field. This research brings together evidence, pointing to the need for a paradigm shift in the food, energy, and water nexus toward greater policy integration. The use of bioenergy may promote energy security, revitalize rural economies, and reduce greenhouse gas emissions if implemented at the local community level. However, meeting food and energy needs in the twenty-first century, especially in Southeast Asia, will require a range of approaches to shift the current development paradigm in order to enhance natural resource governance and climate-friendly agricultural practices. This research shows that if current practices prevail, Thailand is likely to use an unsustainable model of trade-led development

This chapter is written in homage to Ulrich Beck and Daniel Maiello, whose lives inspired ideas that changed the way we think about nature and the way we choose to live. A version of this chapter was presented at the Water-Energy-Food Roundtable in Brasilia in May 2014. The authors gratefully acknowledge the contribution of Chatthip Nartsupha, Charit Tingsabadh, Wanun Pempibul, Nopadol Suanprasert, Chumroen Benchavitvilai, Thongchanh Sengsoulivong, Athiras and Witchuda Dumdee, Andrew Bartlett and the LaoFAB network, Ursula Shaw, Daniel Maiello, Máximo Baiardi, and, especially, Francisco Cannabrava. Parts of the background research were funded by the Food and Agriculture Organization (FAO) of the United Nations and the International Institute for Sustainable Development (IISD).

S. Shaw (✉)
International Institute for Sustainable Development (IISD), Buenos Aires, Argentina
e-mail: Sabrina@iisd.org

S. Jai-In
Alternative Energy Foundation, Institute of Thailand (AEF), Bangkok, Thailand
e-mail: samai22@gmail.com

M. Juntopas
Climate Change Adaptation Specialist, Asian Development Bank (ADB), Bangkok, Thailand
e-mail: muanpong101@gmail.com

M. Lopez and J. Suryomenggolo (eds.), *Environmental Resources Use and Challenges in Contemporary Southeast Asia*, Asia in Transition 7,
https://doi.org/10.1007/978-981-10-8881-0_3

at the regional level. To the end, this chapter puts forward an alternative development strategy to build on the synergies between small-scale initiatives and agro-energy sustainability. However, evidence indicates that these synergies will be difficult to capture in the current policy climate that separates people's well-being, social progress, and ecosystem sustainability from intensifying resource use in the Mekong.

Keywords Mekong region · Thailand · Trade · Regional integration
Bioenergy · Agro-energy · Biofuels · Sustainable development
Resource use · Community-based resource management · Smallholders
Value chains · Climate change

Introduction

Trade between Thailand and its Mekong neighbors has shown an unprecedented level of growth in the past few decades.[1] With the entry into force of the ASEAN Economic Community in 2016, enhanced market integration will give further impulsion to regional systems of food, fuel, and fiber. This regional integration will not be without social and environmental impacts. Rapidly increasing demand for energy has led to the expansion of first-generation bioenergy produced from agricultural biomass. Such a trend, if not managed, has the potential to displace food production and impact the food supply chain (Greater Mekong Forum on Water, Food & Energy 2015; Giampietro and Mayumi 2009; FAO 2008; UNE 2007; Doornbosch and Steenblik 2007).

A review of the literature makes it clear that the current energy situation in the Mekong is not sustainable due to a rapidly increasing demand from growing economies and reliance on energy imports; this situation has been made more complex given the uncertainty of climate change impacts (Bruckman et al. 2016; WWF 2016). In this context, bioenergy is considered to be one way to lessen the region's dependence on oil imports and diversify its energy mix, while at the same time stimulating regional connectivity (Bruckman et al. 2016; Smajgl and Ward 2013; Tharakan et al. 2012; ADB 2009a).

Trade in bioenergy represents opportunities as well as challenges for the region. This is particularly the case for Thailand as a leading global food supplier and the eighth largest agro-energy producer. In order to capture the opportunities, bioenergy strategies need to bring together theory, policy, and practice in a way that allows local experiences to inform policymaking. Hence, the approach being pursued

[1]The Mekong is comprised of Cambodia, Lao PDR, China (Yunnan and Guangxi), Myanmar, Thailand, and Vietnam.

should ensure agro-crops contribute to sustainable development and do not replace food crops or result in forest clearing or biodiversity loss. The expectation is that the coming decades will witness a transition to more resource-efficient generations of bioenergy (e.g., algae, cellulosic) (Samai 2016; USDA 2016; ADB 2013; UNEP 2011).

Thailand has gained experience in adapting to the impacts of globalization, which have tested traditional ways of structuring the economy. Importantly, the *unintended* environmental consequences of modernity have emerged as a *common threat* to our existence (Beck 1992). Modernization of the Thai economy has included strengthening community-based agro-energy initiatives. In Thailand, small-scale farmers predominately cultivate plots of between 2–5 hectares. The Thai model for integration of smallholders into the agro-energy value chain offers a basis from which to build resilience at the local level. This chapter offers evidence that this model can also help to bring about a transition to low-carbon economies in a highly climate-vulnerable region.

Methodology: Goal, Scope, and Context of the Research

This chapter explores the relationship between Thailand's trade in bioenergy based on evidence from the field. The shift to bioenergy was launched in Thailand's 15-year Renewable Energy Development Plan in 2008 and strengthened in the Alternative Energy Development Plan (AEDP) in 2015. The AEDP now forms a part of Thailand's Integrated Energy Blueprint (2015–2036), endorsed in October 2015 alongside the National Economic and Social Development Plan. The AEDP sets targets to increase alternative energy consumption to 30% by 2036, with ethanol projected to increase from 1.17 billion liters (BL) in 2015 to 2.6 BL by 2036, and biodiesel from 1.23 BL in 2015 to 5.1 BL by 2036. The expectation is for second-generation bioenergy to increase, with cellulosic technology (USDA 2016; IEA 2015, 2011; ADB 2009a).[2] Data for this research are based on a literature review, fieldwork using a survey questionnaire, observation, and semi-structured interviews with 143 respondents between 2008 and 2010.

Table 3.1 contains the list of informant organizations by category. While the focus was Thailand, the entire region is covered to some extent. The survey questionnaire was designed to gain insight into the challenges and opportunities presented by agro-energy development in Thailand and the Mekong as perceived by the representatives of the five informant categories. The questionnaire consisted of ten questions with multiple choice answers (Table 3.2).

[2]As in most countries, fossil fuels continue to be heavily subsidized. In 2012, Thailand spent US$7 billion on fossil fuel subsidies, amounting to 1.9% of GDP (IISD-GSI 2015).

Table 3.1 Informants by category and method of data collection

Method of data collection	C1 Government		C2 International & regional intergov organizations		C3 NGOs		C4 Private sector		C5 Local level		TOTAL	
Informant Interviews	114		110		173		36		76		545	
In-depth informant interviews[a]	29		25		38		28		22		143	
Questionnaire	29		25		38		28		22		143	
Focus groups[b] individual/mixed	6	14	4	6	4	17	11	6	3	7	21	28

[a]Informants for the in-depth interviews also responded to the survey questionnaire

[b]Of the total of 49 focus groups, 28 were amongst individual categories of informants and 21 were mixed to bring together informants from different categories

Table 3.2 Survey questionnaire

1. What is the main driving force behind your country's *policies* on biofuels?
2. Could your Government benefit from more **information** in formulating biofuels policies?
3. Could **coordination** between ministries on biofuels policy be *improved* in your country?
4. What is the *main challenge* to the development of the **biodiesel** sector in your country?
5. What is the *main challenge* to the development of the **ethanol** sector in your country?
6. Which actor should *take the lead* in developing the biofuels sector **in your country**?
7. Which actor should *take the lead* in developing the biofuels sector **in the Mekong region**?
8. Are you aware of, or do you benefit from **Government incentives** to promote biofuels?
9. Are you aware of, or do you make use of **standards** related to biofuels?
10. In your opinion, is your country's biofuels sector *developing* along the **right path**?

Interviews were conducted with five categories of key actors in the bioenergy sector. This was supplemented with focus group discussion. The objectives of the field survey were the following:

(a) To assess the challenges and opportunities related to Thailand's development of bioenergy.[3]
(b) To investigate how a broad range of actors is integrating trade and environment considerations related to bioenergy.
(c) To assess the implications of bioenergy development on sustainable development in the context of the Mekong region.

Field research was conducted through six case studies, as outlined in Table 3.3, based on the recommendations of experts in the field. The first set of three case studies focuses on community bioenergy in different regions in Thailand. The second set of three case studies illustrates commercial bioenergy operations in selected sites in Lao PDR, Myanmar, and Thailand.

In order to assess the policy options for bioenergy development in Thailand and the Mekong region, the investigation distinguishes between the following three categories of bioenergy development:

Type A: community biodiesel to revitalize rural development and build resilience in food and energy systems at the *local* level (in case studies 1, 2, and 3).

Type B: commercial ethanol and biodiesel production for *domestic* use, with the potential to export tapioca-based ethanol (in case study 4).

Type C: Thai investment in agricultural feedstock in the neighboring Mekong countries (Cambodia, Lao PDR, and Myanmar) for export to Thailand to be used in commercial ethanol and biodiesel production (in case studies 5 and 6).

[3]Bioenergy includes liquid, solid or gaseous fuels produced from plant biomass, such as from agricultural crops and by-products, aquatic plants, forestry products, wastes and residues, and animal wastes. Liquid bioenergy, the subject of this chapter, include biodiesel (from oilseeds such as palm oil or *jatropha curcas*) and ethanol (from sugar, cassava, maize, and other starchy crops).

Table 3.3 Case studies of community and commercial bioenergy

	Agro-energy crop	Cultivation area
PART I: Community biodiesel		
Case study 1	Oil palm	Rangsit, **Central Thailand**
Case study 2	Oil nuts (Jatropha)	Vanghinlad, Chumpae, **Northeastern Thailand**
Case study 3	Oil palm	Aoluk, Krabi, **Southern Thailand**
PART II: Commercial agro-energy		
Case study 4	Sugarcane and cassava	Khon Kaen, **Northeastern Thailand**
Case study 5	Oil palm and cassava intercropping; Jatropha	Pakse, Champassak, **Southwestern Lao PDR**
Case study 6	Oil nuts (Jatropha)	Shan state and Mandalay, **Myanmar**

This chapter puts forward that despite the policy frameworks in the Ayeyawady-Chao Phraya-Mekong Economic Cooperation Strategy (ACMECS) and ASEAN, regional governance of the agricultural sector—land, water, trade, and investment policies—is fundamentally *lacking in practice* in the Mekong. That is why an alternative policy approach, the *Small is Smart* option, is advanced to integrate small-scale farmers into a broader framework for agro-energy production and use. It envisages smallholders as spearheading sustainable resource management and tackling energy poverty. Moreover, it envisages the incorporation of smallholder efforts into a coordinated regional agro-energy supply chain.

Thailand's Bioenergy Strategies: Why Bioenergy?

Economic growth is propelling significant increases in energy production, primarily based on conventional sources (coal, natural gas, oil, and large-scale hydropower). It is estimated that energy demands in the Mekong will triple by 2030 (ADB 2015). As the world's largest exporter of tapioca and rice, and second largest sugar exporter, Thailand has significant capacity to produce feedstocks to generate bioenergy (Chanthawong and Dhakal 2016; Samai 2016).

Thailand's bioenergy sector is more advanced than other countries in Asia as a whole for several reasons. First, Thailand's dependence on petroleum imports to stimulate its export-led growth has provided an incentive to develop alternative energy sources. Over the past several decades, Thailand has spent approximately 10% of its annual gross domestic product (GDP) on oil imports. Already in the 1980s, Thailand began investing in research and development to decrease oil dependence by converting its abundant agricultural biomass into biofuels

(Achawangkul 2015). Thailand's energy intensity in relation to GDP has been rising since the early 1980s to a relatively high level. For each percent increase in GDP, there is a resulting increase of 1.4% in energy consumption. The majority of energy is consumed in the industrial (37.1%) and transport (35.4%) sectors (DEDE 2015b). The consequences of these dramatic increases will continue to be unprecedented in terms of regional livelihoods, ecosystems, and greenhouse gas emissions (WWF 2016; Greater Mekong Forum on Water, Food & Energy 2015).

Second, in light of the volatile world price of oil, bioenergy offer a means to diversify fuel sources, thereby increasing energy security (Bruckman et al. 2016; Achawangkul 2015; Apichart 2015; Tharakan et al. 2012; ADB 2009b, c, d; Bundit 2009). In this respect, the vast majority of respondents (82%) in the research survey ranked energy security as the main driving force behind bioenergy development in the Mekong, with only one respondent referring to climate change as a driver for the development of the sector.

Third, developing the bioenergy sector represents an opportunity to add value to the agricultural sector and stimulate rural development. Smallholders have the possibility to meet their energy needs and to contribute to the agricultural value chain (Tharakan et al. 2012; FAO 2012b).

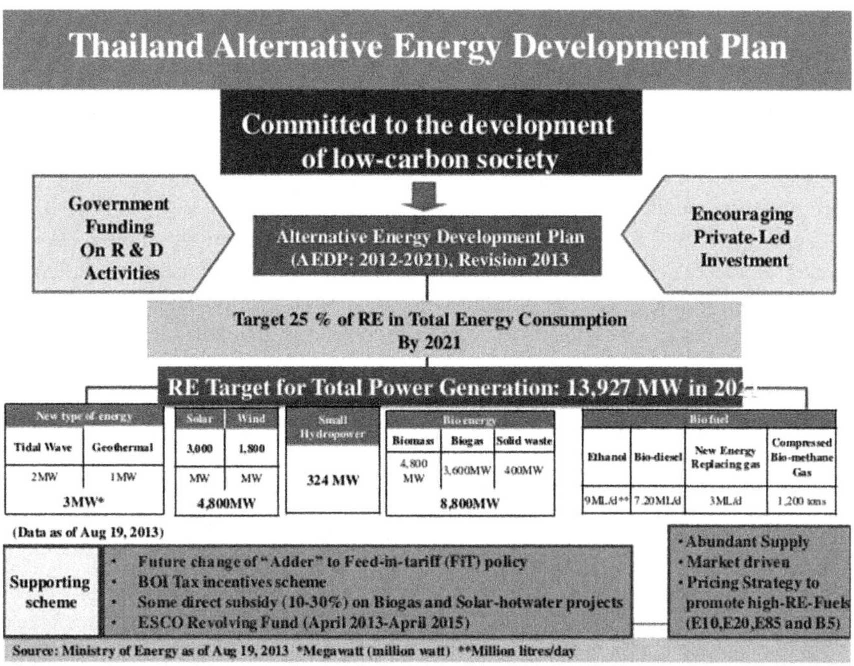

Fig. 3.1 Thailand's Alternative Energy Development Plan. *Source* Ministry of Energy DEDE (2013)

Fourth, bioenergy may help to reduce greenhouse gas emissions, thereby contributing to addressing climate change. Studies are ongoing to calculate the net energy balance of various biofuel feedstocks (ECOFYS 2016; Johnson and Seebaluck 2012; Guariguanta et al. 2011; ADB 2009c; FAO 2009b; SEI 2008). More efficient use of agro-processing waste materials and water effluents also is reforming the sustainability of the agricultural sector (e.g., for tapioca and oil palm processing mills) (Sombilla et al. 2009; Segschneider 2008; Surapong 2008).

Based on this combination of diverse policy objectives, Thailand's ambitious Alternative Energy Development Plan (2012–2021) includes targets for ethanol of 9 million liters/day and for biodiesel of 5.97 million liters/day by 2021. The mandate for B5, instituted in 2012, was increased to B7 (provisionally decreased to B4 in 2014). As outlined in the Fig. 3.1, Thailand's goal is to supply 25% of national energy consumption with renewable energy by 2021 (compared with 9.4% in 2015) (DEDE 2015a).

Prospects for Trade in Biofuels

Thailand aims to increase the production of bioenergy primarily to meet domestic targets in the transport sector. Transport fuel currently blends ethanol with gasoline in two ways: E20 (a blend of 20% ethanol and 80% gasoline) and E85 (a blend of 85% ethanol and 15% gasoline). Over the past decade, its strategic plans for gasohol and biodiesel have established blending targets. Driving these targets are plans to increase sugar plantations from 10 million rai in 2015 to 16 million rai by 2026, with the intention of encouraging rice farmers to switch to higher value crops, such as sugarcane and oil palm.[4]

The overall aim is to diversify Thailand's renewable energy matrix. In the context of increasing economic integration, rising regional investment and mandated renewable energy targets in other countries, come opportunities for bioenergy trade with Mekong countries, ASEAN, and beyond. Diversifying cropland for bioenergy becomes appealing. At the same time, there is concern about the environmental and social consequences of increasing the contribution of agricultural feedstocks for use as fuel. In particular, concerns about the impact on food security have been strongly voiced (Fullbrook 2013; UNESCAP 2009; FAO 2008). On the other hand, there is research indicating that use of bioenergy—under the appropriate conditions, may generate economic, environmental, and social benefits (Tharakan et al. 2012; Johnson and Seebaluck 2012; FAO 2012b).

Over the past few decades, Thailand has increased significantly its production, consumption and export of ethanol. More recently, there has been a slow but steady

[4]The Thai Government has recently approved 25 licenses to boost sugar factories raising them to 79 in total. See *25 New licences Sweeten Sugar Output,* Bangkok Post, 7 Sept, 2016. http://www.bangkokpost.com/business/news/1080080/25-new-licences-sweeten-sugar-output Accessed 15 Sept, 2016.

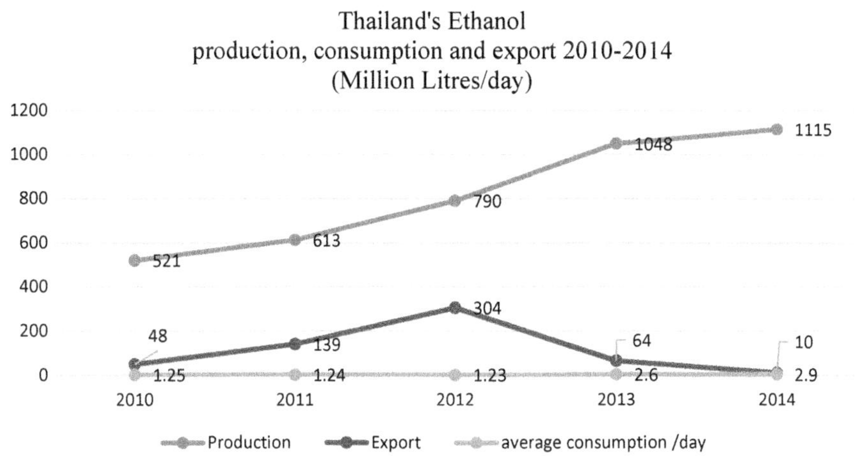

Fig. 3.2 Thailand's ethanol production, consumption, and export (2010–2014). (Million liters per day) *Source* DEDE Ministry of Energy (2015a)

increase in the use of ethanol in the domestic fuel mix in the transportation sector. Correspondingly, more of what is produced is used domestically. Figure 3.2 shows these three trends.

Only 21 of the 47 licensed ethanol plants are in operation, with a capacity of 3.7 million liters per day (see Fig. 3.4). Figure 3.3 sets out the location of the plants and the clusters of distribution. Notably, these clusters are in close proximity to borders with Cambodia, Lao PDR, and Myanmar, as well as Malaysia to the south, positioning Thailand as a bioenergy hub for the region (Samai 2016; Chantankome 2016). In 2016, the government approved licenses for 25 more ethanol plants, bringing the total to 79 plants by 2021, with a capacity of 5.4 million liters per day (Bangkok Post 2016).

Since 2014, exports of ethanol have declined significantly mainly due to increasing domestic demand. In 2014, Thailand exported only 8 million liters, primarily to the Philippines. In contrast, Thailand exported 167 million liters of ethanol in 2011 and 303 million liters in 2012 to a range of countries, including Australia, Japan, the Netherlands, the Philippines, Singapore, and Taiwan (See Fig. 3.5; DEDE 2015a).

Ethanol exports are constrained by the fact that export licenses must be secured from at least three government agencies. In addition, the Cane and Sugar Act (1984) does not differentiate sufficiently between alcohol production for beverage use and for fuel.[5] The Thai Ethanol Manufacturers Association argues that the regulatory framework should be revised to reflect this distinction to facilitate exports. Moreover, it would increase the economic viability of the sector to permit ethanol to

[5]The categorization of ethanol as an agricultural product or an industrial good is an issue that also needs to be clarified in the World Trade Organization.

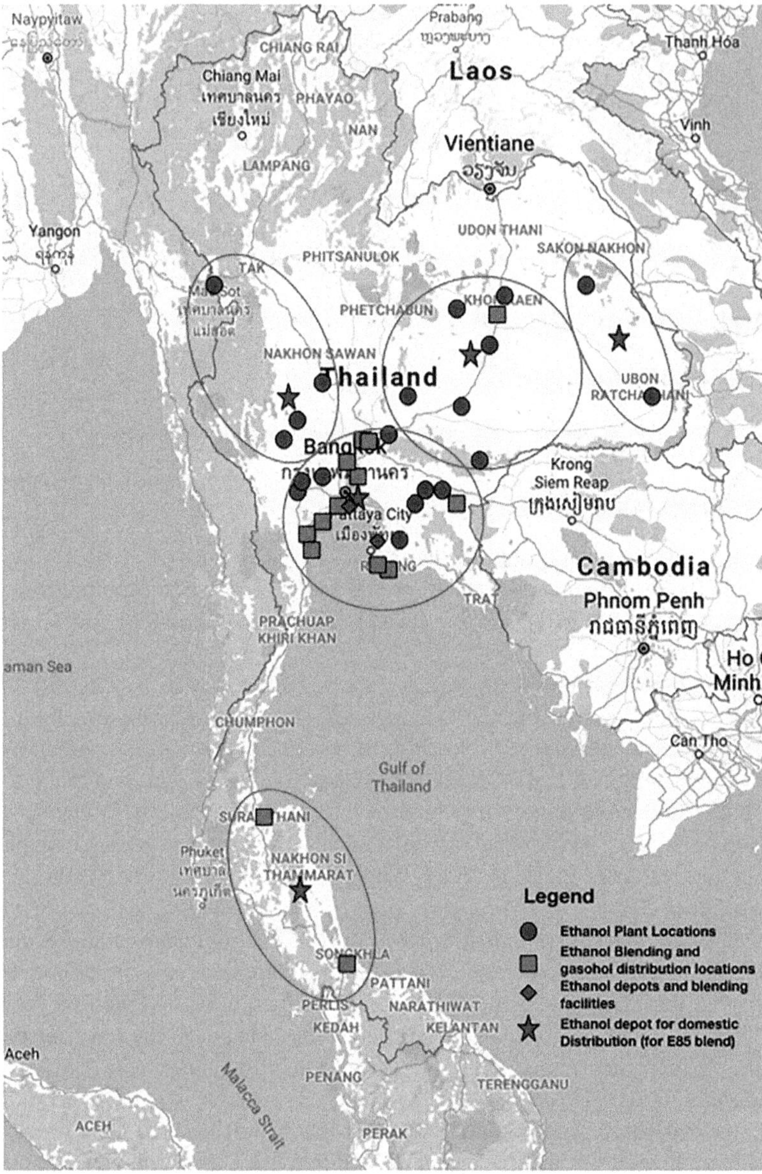

Fig. 3.3 Clustering of ethanol distribution in Thailand. *Source* Chantankome (2016)

be produced directly from sugarcane juice (as opposed to molasses) and allowing market dynamics to drive the degree of substitution between sugar and ethanol production. In this respect, the majority of respondents (68%) in the research survey considered that the main challenge to the development of the ethanol sector is a

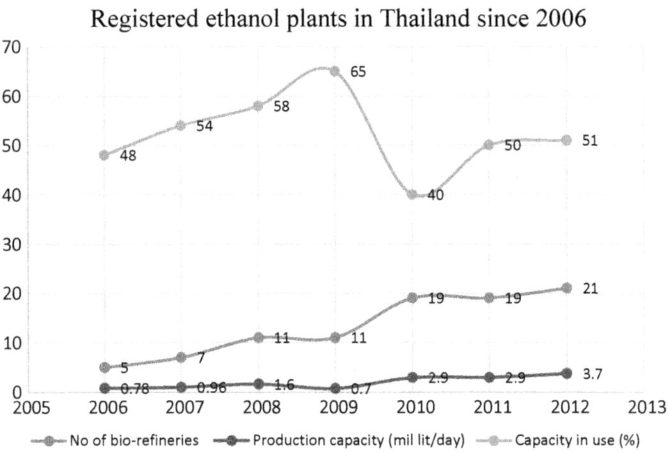

Fig. 3.4 Thailand's registered ethanol plants since 2006. *Source* DEDE (2015a); Kumar et al. (2013)

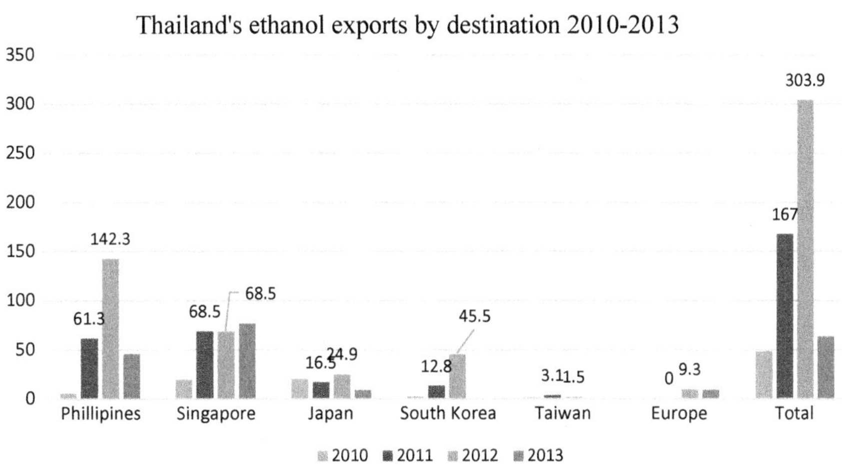

Fig. 3.5 Thailand's ethanol exports by destination 2010–2013. *Source* DEDE (2015a)

combination of a consistent policy framework and policy implementation, as opposed to improving crop yields, harvesting techniques or processing technologies.

In contrast to the scenario for the ethanol sector, there is insufficient domestic supply of *crude palm oil* to meet the national target. The target is to increase biodiesel production capacity to meet the target of 1.25 billion liters of B100 in 2016. There are 10 producers in operation in registered biodiesel plants with an estimated total production capacity of 5.4 million liters per day (1.63 billion liters per year) (Fig. 3.6) (Samai 2016). Exports of biodiesel amounted to 1,870 metric

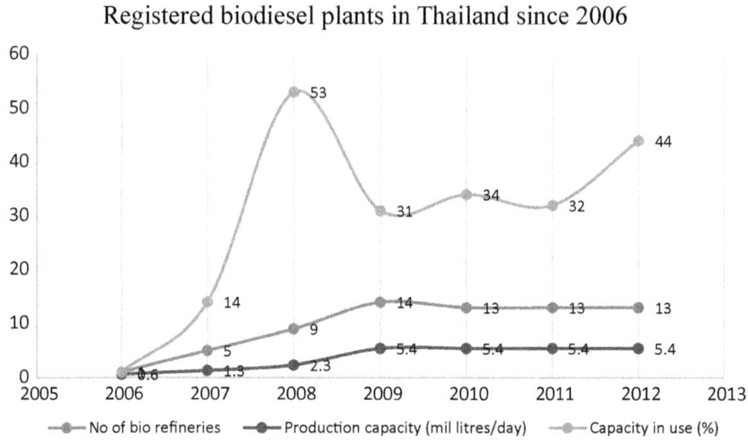

Fig. 3.6 Registered biodiesel plants in Thailand since 2006. *Source* DEDE (2015a)

tons in 2014, with imports reaching 2,810 metric tons. The target is to reach 7.2 million liters per day by 2021.

Since 2007, it has been mandatory in Thailand to blend diesel initially with 2% biodiesel and 98% diesel (so-called B2) and increasing, incrementally, through B3, B4, and B5 to reach B7 as of January 2016 (with fluctuations) (Samai 2016). To meet these blending requirements, Thailand has put in place a series of plans to increase the area of oil palm plantations to 5.5 million rai (880,000 ha) by 2021. Average yields are expected to reach 3.2 million tons per rai (30 MT/ha) by 2021, with crude palm oil crushing rates attaining over 18%. Thailand's planned acreage and production of palm oil indicate its domestic consumption in the food sector and potential for exports. Noncommercial, small-scale biodiesel production and use are also being promoted in nearly 500 communities to enhance local energy sufficiency.

The majority of respondents (72%) in the research survey considered that the main challenge to biodiesel development is a combination of a consistent policy framework and policy implementation, as opposed to the need to improve crop yields, harvesting techniques or processing technologies. The government is supporting biodiesel R&D to promote energy crops such as *jatropha curcas* and microalgae, diesohol (blending of ethanol and diesel), and oil conversion technology, such as bio hydrofined diesel (BHD) and biomass to liquid (BTL).[6] The target for commercial production of these sources is 2 million liters per day by 2018, reaching 25 million liters per day by 2021 (Samai 2016).

There are four key points related to trade in bioenergy that must be kept in mind. First, Thailand's expansion of bioenergy capacity initially needs to be supported by

[6]Thai Oleochemicals, a subsidiary of PTT, introduced a BHD product on the market in 2013, with total sales of bio hydrofined diesel reaching 50,000–80,000 L per day.

a policy and regulatory framework that establishes a secure domestic market (Shaw 2009). Second, the removal of regulatory barriers to trade would facilitate exports in the region. Third, the development of high-quality performance standards[7] for biofuels at the regional level would encourage an integrated bioenergy market in the Mekong. Since 2008, Thailand established dual product quality standards for biodiesel (community and commercial) and another for ethanol. The Roundtable for Sustainable Biomaterials (RSB) and the Roundtable for Sustainable Palm Oil (RSPO) are among the standards bodies developing sustainability indicators for biofuels (RSPO 2016; RSB 2011).

Fourth, Thailand's main trading partners, the European Union, the United States, and Japan are enacting strict sustainability regulations for biofuel imports. This means that any future development of trade in biofuels may depend on certifying the sustainability of the supply chain for these different regulatory regimes. Ethanol producers are motivated to certify for sustainable practices by the demand for certified ethanol for land transport in the EU market (Potts et al. 2014).[8] A controversial element of the EU policy mandates the use of biofuels lead to a 35% saving of greenhouse gas emissions calculated during the life cycle of the project.[9] In 2015, the EU legislation on biofuels was amended to address the risk of indirect land use change and to prepare the transition to advanced biofuels (EC 2015).

Developing Energy Crops in the Mekong

Developing the bioenergy sector will have a significant impact for Thailand as well as the predominantly agriculture-based Mekong economies as a region. Agriculture serves to underpin rural incomes, food supply, and increasingly, feedstocks for the expanding bioenergy sector. The contribution of the agricultural sector to GDP and exports in Thailand has decreased significantly since the mid-1980s as the labor force has been employed in the industrial and services sectors. Nevertheless, the contribution of agriculture to employment remains at 40% (Fig. 3.7).

While the contribution of the agricultural sector has been declining in Thailand, it continues to contribute 55–75% in other Mekong countries. Consequently, the

[7]Standards depend on the crop and are set by the Roundtable for Sustainable Biofuels (RSB), RSPO and Bonsucro.

[8]The EU Directive 2009/28/EC (adopted in 2003 and revised in 2009) sets out that 10% of the transport fuel of each member country come from renewable sources such as biofuels by 2020. Fuel suppliers are also required to reduce the greenhouse gas intensity of the EU fuel mix by 6% by 2020 in comparison to 2010 (EC 2009).

[9]The EU has outlined a set of sustainability criteria to ensure that the use of biofuels (in transport) and bioliquids (for electricity) provides carbon savings and protects biodiversity in fulfiling renewable energy targets. To be considered sustainable, biofuels must achieve greenhouse gas savings of at least 35% throughout the life cycle in comparison to fossil fuels, rising to 50% in 2017 and to 60% in 2018 for new production plants.

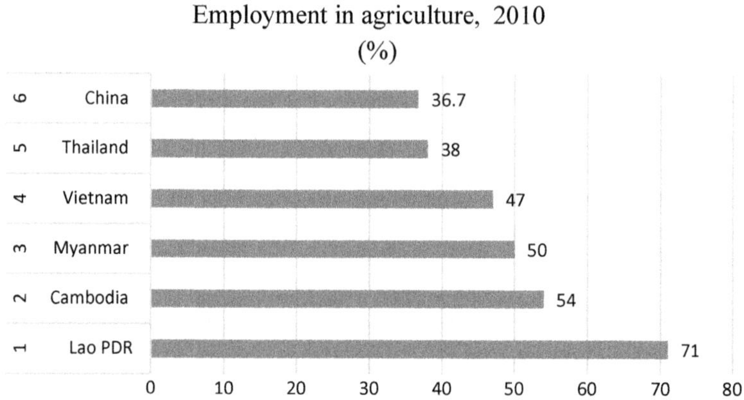

Fig. 3.7 Employment in agriculture in Mekong countries, 2010 (%). *Source* World Bank database available at http://data.worldbank.org/indicator/SL.AGR.EMPL.ZS Vietnam 2012 data; China: http://www.statista.com/statistics/270327/distribution-of-the-workforce-across-economic-sectors-in-china/

input of agriculture to GDP and exports is also declining. In 2014, agriculture contributed 8.4% to Thailand's GDP and 8% to exports. By contrast, agriculture continues to contribute between 20% and 45% to the GDP of other Mekong economies and remains responsible for a significant contribution to domestic food supply.

As illustrated in Fig. 3.8, Thailand's annual expenditure on fuel imports has been rising exponentially relative to agricultural exports since the turn of the century. There are opportunities to add value to agricultural production through developing energy crops. In a region, in which a significant majority of the population still works in the field, adding value to the agricultural value chain is recognized as vital to improving sustainable livelihoods. In Thailand, for example, the vast majority of oil palm growers are independent smallholders, in contrast to the large-scale plantations in Malaysia and Indonesia. Nevertheless, several

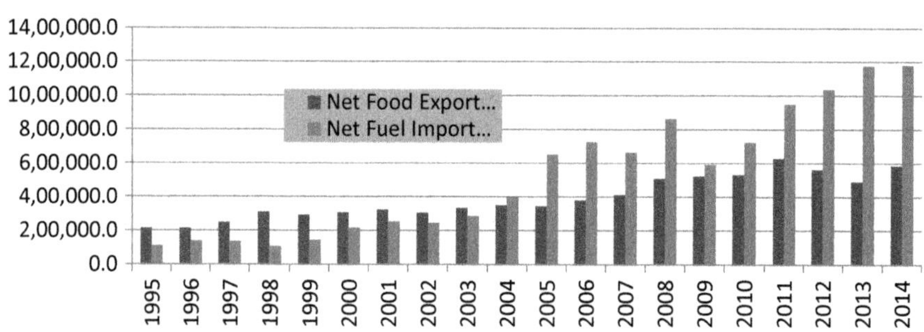

Fig. 3.8 Fuel Imports and Agricultural Exports, 1995–2014. *Source* Samai (2016) based on latest data from the Thai National Statistics Office and the Ministry of Energy

challenges to agro-energy development are identified in the literature, primarily related to food security and environmental sustainability (Tharakan et al. 2012; Pimental 2009; FAO 2008; World Bank 2008).

With countries in the Mekong region considering renewable energy targets, investment in bioenergy is increasing at a rapid pace in response to target mandates set out in regulatory regimes (Table 3.4). Notwithstanding these efforts, policies promoting sustainable bioenergy are perceived to be lacking, with all respondents of the survey undertaken for this research overwhelmingly indicating that information and coordination need to be improved.

There are a multitude of initiatives underway to promote agro-energy cooperation in the Mekong, such as the ACMECS and the ASEAN Free Trade Area (AFTA). In the context of the latter, Thailand is providing assistance for energy crop contract farming along border areas with Lao PDR, Cambodia, and Myanmar and to allow tariff-free imports of certain agricultural products. This has led some experts to contend that the current contract-farming model for agro-energy has shifted the paradigm in agricultural development in the Mekong region (Fullbrook 2007). For Thailand, contract farming in neighboring countries offers a means to effectively expand agricultural production. It is expected to contribute 1 million rais (160,000 ha) to oil palm acreage in the region by 2020. As the second largest ethanol producer after China in the Mekong, Thailand is likely to increase production and exports of ethanol with the liberalization of trade under AFTA, potentially relying on feedstock from neighboring countries to meet blending mandates.

Exploring the Policy Options: Energy, Food Security, and Climate Adaptation

Strategies to increase the contribution of bioenergy to Thailand's energy mix have to take into account several related issues, notably the consequences for food security and the environment. First, Thailand has initiated a coordinated national debate on how to balance food and fuel requirements from agriculture. To ensure that a rising demand for energy crops does not negatively impact production for food, the agricultural sector is being zoned and restructured (Samai 2016).

Second, given Thailand's position among the world's leading producers and exporters of rice, sugar, and tapioca, the issue of food security acquires an international dimension and Thailand's production not only impacts domestic supply, but also the global food supply chain (Yang et al. 2009). What is crucial here is that a switch from food to fuel cultivation, if not properly managed, would adversely impact global food supply (Fullbrook 2013; Shaw 2010). In this respect, there is a growing body of evidence examining the impact on food security and the socio-environmental impacts of agro-energy. A study by Silalertruksa and Gheewala (2012) on ethanol in Thailand confirms that greenhouse gas

Table 3.4 National Bioenergy Policies in the Mekong Region

Country	Agro-energy crops		Targets in transportation (% share of total fuel consumption)		Policy and fiscal incentives	Import/Export Tariff rates
	Biodiesel	Ethanol	Biodiesel	Ethanol		
Cambodia	Oil palm Jatropha	Cassava Sugarcane	*Proposed 5% by 2016*		Tax exemptions; investment incentives; Biodiesel Development Fund	Benefits from General System of Preferences (GSP) and EU Everything But Arms Initiative
China	Jatropha	Corn Wheat Sorghum Cassava Sugarcane	10% by 2020 (Renewable Energy Law) ethanol 12 million Liters by 2020 Biodiesel 6 million Liters by 2020		Tax exemptions; blending credits; low-interest loans; R&D funding; ethanol production subsidies	Biodiesel applied tariff 9% Denatured ethanol: ad valorem 80% Undenatured ethanol: 100%
Lao PDR	Oil palm Jatropha	Sugarcane Cassava	E10 by 2015 E20 by 2020		Investment promotion	GSP and EU Everything But Arms Initiative
Myanmar	Jatropha Oil palm Soybean Sesame	Sugarcane Sorghum Cassava Maize, potato	E5 and E15 B5 to B20		Demonstration projects	Benefits from GSP and EU Everything But Arms Initiative
Thailand	Oil palm Jatropha Cooking oil	Sugarcane Cassava	B5 mandated since 2001 1,643 million Liters by 2022	Gasohol (E20) available nationwide 3,285 million Liters by 2022	Price guarantees; blending credits; tax incentives for gasohol vehicles; R&D funding; investment promotion	Tariff rate quota regime for imports of Crude Palm Oil Biodiesel: ad valorem 5% Denatured ethanol: 21-22 Baht/L
Vietnam	Fish oil Jatropha Cooking oil	Sugarcane Sorghum Cassava	E5 128 million Liters by 2020	5% of total fuel demand in transport sector by 2020 684 million Liters by 2020	R&D funding; investment promotion	Benefits from GSP and EU Everything But Arms Initiative

Source Compiled based on Samai (2016); DEDE (2015a); USDA (2016); Thakaran et al. (2012)

(GHG) emissions depend on the management of crop residues and, especially if there is direct land use change and conversion of tropical forest to cropland, can result in a loss of biomass and increased CO_2 emissions. Silalertruksa et al. (2012a) assess the impacts of indirect land use change (ILUC) of Thai ethanol production to find that the displacement of the cultivated area of other crops (sugarcane) could result in a larger impact on GHG emissions mainly due to the potential change in biomass and soil carbon stock.

Agro-energy production can also be expected to have significant impacts on water resources (UNEP 2011). Given that the water footprint is sensitive to location, Kumar et al. (2013) emphasize the importance of local studies to construct a larger picture of sustainability impacts. To this end, a study of the water footprint of biofuels in the Khlong Phlo watershed in central Thailand found that production and land use change would impact water quality (Babel et al. 2011). Studies are also ongoing to examine the socioeconomic impacts of the AEDP targets for ethanol and biodiesel, some forecasting that bioenergy can contribute to employment and GDP (Malik et al. 2009), while others signal concerns over elements of sustainability (Silalerstruksa et al. 2012a; Salvatore and Damen 2010).

Third, and consequently, one of the reasons why Thailand has recognized the urgency to address climate change is its potential impact on agricultural productivity. Agricultural systems are dependent on imported fossil fuels and are vulnerable to climate change. The need to adapt to climate change has strengthened Thailand's projects for community bioenergy. Fourth, the Clean Development Mechanism (CDM) of the Kyoto Protocol is stimulating renewable energy projects. As of 2016, Thailand has approved 280 CDM projects, including 110 for biogas operations (mainly from tapioca and oil palm processing wastewater) and 60 for biomass operations (mainly from rice husks) (TGO 2016; ECOFYS 2016). The increase in CDM projects is driving greater awareness since the survey was conducted for this research, in which few respondents (1%) indicated that climate change mitigation was a driving force for bioenergy development in the Mekong. Further research is necessary to determine the extent to which this profile has changed in the region.

Opportunities and Challenges for Thailand's Bioenergy Sector

Finding the appropriate balance between energy security, energy efficiency, and agro-energy trade and development presents challenges as well as opportunities (Table 3.5), which require policy coordination, technological and productivity innovation and sustainable resource management. The impacts of natural resource-led development in the Mekong increasingly are being explored as interlinked systems of water, food, and energy (Greater Mekong Forum on Water,

Table 3.5 Prospects for Trade in Bioenergy in the Mekong

Opportunities	Challenges
To add value to agricultural production; to use efficiently agricultural waste residues	Need for well-defined bioenergy policies and targets at the national and regional level
To shift dependence from petroleum to enhance energy security	Need to address food security concerns and balance competing claims for land
To gain from the Mekong's diversity of energy crops for ethanol and biodiesel	Need for economic incentives to deliver sufficient feedstocks; need to enhance awareness and information on the bioenergy market
To develop small-scale community bioenergy production and use to build resilience to external shocks; create local energy sufficiency; stimulate rural development and reduce input costs from petroleum and fertilizers	Need to balance large-scale commercialized agro-industrial development of bioenergy for the transport sector and for export
To make use of low cost of production (land, labor, and water)	Need to address land use changes and labor migration within the region resulting from bioenergy development; need to ensure transparent, equitable investment and safeguard land rights
To develop infrastructure in a dynamic region	Need to enforce strategic impact assessments to address social and environmental consequences of bioenergy plans and projects
To operationalize the Clean Development Mechanism (CDM), gain carbon credits, and stimulate investment and technological innovation in second-generation bioenergy	Need to guide private sector investment through enforcing socio-environmental regulations (air, water, land, and labor laws)
To develop the potential to export bioenergy production	Need to facilitate exports and secure market access and meet sustainability criteria

Source Compiled by the authors

Food & Energy 2015; Foran 2013; Smajgl and Ward 2013) to address resource scarcity (Allouche et al. 2015).

In large part, due to the fact that demand for bioenergy is driven by regulatory mandates, with production costs subsidized by governments, there are valid concerns about economic efficiency and socio-environmental sustainability (IISD/GSI 2012). Reports on the biofuels sector in various countries highlight the risks inherent in subsidizing fuel: subsidies increase consumption and discourage more efficient resource use (IISD/GSI 2015; ADB 2015; Lopez and Laan 2008; Steenblik 2007). Moreover, the impacts of converting 'marginal lands' to feedstock production are highly dependent on local circumstances. Indeed, many aspects of sustainability are context-specific, requiring a complex assessment of the net balance of greenhouse gas emissions due to indirect land use change from shifting to agro-energy crops (what is referred to as ILUC in the climate change discourse) (Tilman et al. 2009).

Water, its management and scarcity are essential elements that affect the future of agro-energy in the Mekong region (ADB 2013; FAO 2011). In the case of sugarcane, for example, most production in Thailand is located in rain-fed areas, with only 10% in irrigated zones (Kumar et al. 2013). Increasing the efficiency of water use to avoid shortages needs to be supported by proper pricing of water to reflect its scarcity.

The focus of international standardizing bodies, such as the ISO, has been to further life cycle analysis of bioenergy crops and sustainability assessments of agro-energy systems. To date, evidence from life cycle assessments in Thailand indicates that there are significant opportunities to adopt new technologies to increase energy efficiency in biofuel conversion and to use agricultural waste residues (Kumar et al. 2013; ADB 2009c; Huang et al. 2009; Nguyen et al. 2008; Jitsanguan 2001). To illustrate the variations in the results of sustainability assessments, depending on the perspective, consider the following. Recognizing that its ambitious 5.75% target for biofuels in the transport sector requires considerable imports, the EU Sustainability Directive stipulates that this target must be met with biofuels that fulfil sustainability criteria. To meet the EU criterion to reduce greenhouse gas (GHG) emissions by 35%, one study examined the GHG emissions of indigenous Irish rapeseed and imported Thai palm oil (Thamsiriroj and Murphy 2009). Given that palm oil generates more biodiesel per hectare than rapeseed and demands less fertilizer and fuel inputs, a reduction in GHG emissions of 29% and 55% were calculated for Irish rapeseed and Thai palm oil systems, respectively. In other words, it appears to be more climate-friendly for Ireland to meet the EU biofuel target by importing palm oil from Thailand.

Transforming Rural Development: Small is Beautiful

Globalization and market liberalization are changing global agricultural production, with the risk of excluding smallholder farmers in developing countries from adding value to their production. The scale of production is a key element in assessing the contribution of bioenergy to sustainable development. Large-scale biofuel systems are facing several challenges that do not paint a picture of sustainability, particularly in developed countries (Giampietro and Mayumi 2009). Moreover, large-scale, energy-intensive mono-cropping plantations are deemed to be one of the main causes of deforestation, soil erosion, and the increased use of chemical fertilizers and pesticides, with impacts on water quality and quantity (FAO 2012a; USAID 2009; Pimental 2006).

Conversely, more efficient use of biomass at the rural level is an attractive alternative to enable a shift to a more sustainable energy matrix. The International Assessment of Agricultural Knowledge, Science and Technology for Development (IAASTD 2009)—a multi-stakeholder United Nations report commissioned by the World Bank and the FAO—critiques conventional industrial agriculture. The report calls for a fundamental change in farming practices to better address increasing food prices, food insecurity, and environmental crises. It reflects a growing consensus among scientists and many governments that the old paradigm of industrial energy and chemical-intensive agriculture is an outdated concept (FAO 2012a). It also points to the role of small-scale farmers and agro-ecological methods in providing a way forward for sustainability in the face of water shortages, soil erosion, and climatic change. The conclusion is that past emphasis on increasing production and yields—the Green Revolution—brought about important benefits, however, these gains occurred at the expense of environment and social equity. The report concludes that more emphasis is required to address the local needs of developing country farmers, including improved access to markets, and infrastructure and financing to integrate into global and regional agricultural value chains.

Small-scale community biodiesel production and use have the potential to empower small landowners with energy sufficiency, thereby lowering their energy input costs and increasing their income (FAO 2012a; Shaw 2010; Altieiri 2009; Penunia 2009; WWF and SNV 2009;FAO 2009a). Such an agricultural transformation positions farmers as frontline environmental stewards with the local knowledge and resilience to contribute to sustainable resource management (Gotsch 2016; Hodbodt and Tomei 2013). In this respect, Thailand's community biodiesel programs have the potential to put into practice local energy sufficiency, whilst simultaneously adding value to the agricultural supply chain and decreasing input expenditures on petroleum and chemical fertilizers and pesticides.

An ADB review (Malik et al. 2009) concludes that biofuels have a significant role to play to meet the Mekong's energy demand, particularly in the transport sector, but that bioenergy expansion will likely impact crop and food prices, both directly and indirectly. Importantly, the extent of the contribution and the environmental impacts will depend on the type of production system pursued. Use of the industrial-scale plantation model would quickly lead to several interrelated social–environmental–political problems that have been observed elsewhere in and outside Asia, namely food versus fuel conflicts, land grab, destruction of forests, and detrimental impacts on soil and water quality (Malik et al. 2009). However, use of a model based on smallholder production, emphasizing nonfood crops and second and third generation technologies will facilitate sustainable bioenergy development in the Mekong. According to Malik and others, this Schumacherian "small is beautiful" approach requires strategic policy interventions and support to the agricultural sector, which have proven to be both controversial and costly (IISD/GSI 2012, 2015).

Market Access and Sustainability Criteria for Land and Water Hungry Crops

Agricultural trade liberalization has always been a critical component of multilateral trade negotiations at the World Trade Organization (WTO). First, commodity-exporting developing countries, in particular, would benefit from greater market access for agricultural products (energy crops) and biofuels (UNCTAD 2008). This would serve to increase revenues for research and development of sustainable production practices and technologies. Second, as a major agricultural exporter, Thailand would gain from agricultural trade liberalization to remove export subsidies in key developed countries, such as Japan, the EU, and the US. As noted in an early Brazilian proposal to the WTO (Brazil 1998), agricultural export subsidies in developed countries distort market access for biofuel exports from developing countries. Notwithstanding price support schemes, for example, for sugar, rice, cassava, and palm oil, Thailand's agricultural sector is considered to be competitive on the world market.

Third, trade trends will be affected by the definition of biofuels as industrial, agricultural or environmental goods. The harmonized system of tariff classification used in the WTO classifies ethanol as an agricultural product with no distinction between its use for fuel or other purposes, whereas biodiesel is classified as an industrial product (Echols 2009: Abdel Motaal 2008; Brazil 2005).

One way to address the social and environmental impacts of bioenergy is to establish and enforce *sustainability criteria.* This is the path forward for biofuels charted by many governments, international organizations, and nongovernmental efforts to address rising concerns (Guariguata et al. 2011). Most criteria to assess sectoral sustainability take into account the socio-environmental effects of *direct* biofuel production, with respect to the land and production processes employed. However, as emphasized by the Dutch Cramer Commission (2007) and World Wide Fund for Nature (WWF) reports (Dehue et al. 2007), the most serious sustainability issues are those related to the *indirect* impacts of large-scale biofuel production, mainly the displacement of other agricultural activities and subsistence farming, as well as changes in land use from forests or grass to crops (Searchinger 2009). Criteria developed to address these impacts by the Roundtables for Sustainable Biofuels (RSB) and Sustainable Palm Oil (RSPO) and other international standards bodies, include lifecycle greenhouse gas emissions, biodiversity, agricultural practices, and social impacts.

Certification of sustainability represents, at the same time, a valuable marketing tool and a costly nontariff barrier to trade, especially for developing countries (Zarrilli and Burnett 2008). It is a tool that is being widely used to address sustainability in ethanol and oil palm biodiesel supply chains in Thailand (Silalertruksa et al. 2012a, b; Silalertruksa and Gheewala 2012). Thailand is by far the leader in the region on environmental standards and regulations. The Thai private sector is adopting environment-related standards developed by the International

Organization for Standardization (e.g., ISO 14000 environmental management standards) to gain market access for its exports. The private sector is also implementing Corporate Social Responsibility (CSR) independently and through Thailand's Business Council for Sustainable Development.

One study on EU–ASEAN trade relations estimates that around 20–25% of EU biofuel consumption by 2020 will be derived from imports. Sustainability criteria agreed by the EU in March 2009 will determine Thailand's market access for palm oil and ethanol exports to the European Union (ECOFYS 2009). In this regard, Argentina, Brazil, Malaysia, and Indonesia are among the key biofuel exporters who have challenged the EU at the WTO.

Sustainability throughout the production of agro-energy and responsible supply chain management will affect trade and have implications, particularly for developing countries (Opijnen and Oldenziel 2011). Compliance with the guidelines developed by voluntary standards-making bodies, such as the Roundtable on Sustainable Biomaterials (RSB) and the Roundtable on Sustainable Palm Oil (RSPO), provide incentives to address socio-environmental impacts (Charnovitz et al. 2008). As with the evolution of eco-labeling and certification over the past several decades, sustainability criteria are deemed to be central and controversial aspects of trade in biofuels (Zarrilli and Burnett 2008). As predominantly voluntary instruments applied to the production process, sustainability certification schemes do not necessarily address macro-level impacts, such as increased food prices and the displacement of food for fuel crops.

Evidence of the sustainability of agro-energy in Thailand is mixed. Since 1990, Thailand's carbon dioxide emissions have increased faster than every other country in the world but one. According to the Global Carbon Atlas (2017), Thailand is the 20th largest carbon polluter in the world (out of 216 countries). A recent review of Thailand's biodiesel prospects concludes that it is arguable whether palm-based biodiesel is economically and environmentally feasible in the long term. This is due primarily to the potential local effects on food supply and prices, as well as changes in land use and agricultural practices related to fertilizer inputs with high-embodied energy costs (Siriwardhana et al. 2009). Moreover, Thailand is developing biofuels to contribute to a domestic demand stimulated by mandated blending requirements for ethanol and biodiesel in the transportation sector. While there is insufficient palm oil to meet domestic demand, exports of ethanol began in 2007. This differs from Malaysia and Indonesia's well-established palm oil capacity predominantly for export. Malaysia and Indonesia account for nearly 90% of global exports of palm oil, primarily to the EU.

Thailand's sugar regime is also adjusting to a new era, with the Sugar Act of 1984 under revision and challenged at the WTO.[10] Moreover, implementation of the ASEAN Economic Community, which came into force in December 2015,

[10]In March 2016, Brazil launched a challenge at the WTO to Thailand's support to sugar producers.

opens up the market, making the outlook for agro-energy in Thailand more complex.

The Way Forward: Results from the Field

Thailand's three-decade experience in managing its sugarcane, cassava, and palm oil industries is worthwhile exploring as it relates to bioenergy to add value to the agricultural production supply chain. The evidence emerging from the survey conducted for this research indicates the need for greater coordination of agro-energy policies in the Mekong to transition to climate-smart agriculture.

In order to achieve the benefits of bioenergy, whilst avoiding potentially harmful consequences, the majority of survey respondents for this research recognized the need to enhance information on bioenergy to enable informed policymaking. This research concludes that bioenergy development in Thailand has the potential to contribute to more efficient use of agricultural biomass. However, there are several constraints from a regional perspective. Agriculture is at a crossroads in the twenty-first century mainly due to mismanagement of trade and environmental policies (IAASTD 2009). It is evident from this research that the increasing demand for agricultural feedstock is placing an unsustainable burden on the natural resource base in the Mekong region. A lack of sufficient political will to coordinate trade and investment policies, particularly in the agricultural sector, continues to be a significant impediment to sustainable resource management. The political panorama may be forced to reform with the likelihood of increasing scarcity of water and land, as Thailand's neighbors open up for the first time to the global economy and regional integration is strengthened.

The hypothesis for this research posited that the prospects for Thailand's trade in bioenergy in the Mekong region are favorable. First, the research supports the conclusion that Thailand's trade in biofuels is likely to increase. Based on an assessment of the current state of play, Thailand is likely to increase exports of ethanol. The main driver of this expansion is Thailand's requirement for blending ethanol and biodiesel in the commercial fuel mix in the transportation sector and increased incentives to feed into the electricity sector. There is likely to be more imports of agro-energy feedstock to meet the blending targets for ethanol. In addition, more imports of palm oil will likely be needed to meet blending requirements for biodiesel, given the lack of sufficient domestic supply of crude palm oil. While domestic consumption at present primarily feeds into the domestic supply of ethanol, trade in the region is likely to grow in ethanol. Moreover, this trade is likely to increase with the commercialization of second-generation biofuels.

Second, based on the current policy framework, the initial hypothesis was put forward that increased trade in bioenergy is likely to have negative implications for sustainable development in Thailand and the Mekong region. The research supports this conclusion as a general point. Nevertheless, a distinction should be made between small-scale bioenergy initiatives (to some extent for tapioca-based ethanol but predominantly for biodiesel) and the commercialized production of biofuels (mainly for sugarcane (molasses)-based ethanol). The conclusion flowing from the field evidence to inform this claim is mixed. The case is cautious in several respects, but generally strong and positive for the development of biofuels in Thailand as supported by the data collected during this research. This is not necessarily the case for the Mekong region as a whole if the current policy framework remains.

On the one hand, the case narratives illustrate the main hypothesis that based on the current policy framework, increased trade in bioenergy is likely to have negative implications for sustainable development in the Mekong region. On the other hand, the outcome could be beneficial if a shift is made toward the *Small is Smart* scenario. As the research suggests, there are bright spots in Thailand that illustrate *Small is Smart*, where smallholder agro-energy is contributing to improving livelihoods, farming practices, and energy efficiency. However, there is insufficient regional coordination of these efforts to bring about a tipping point in renewable energy use or sustainable resource management.

To the extent that an expansion of agro-energy crops displaces forests or biodiversity, the negative consequences of Thailand's current policies are likely to impact more significantly on neighboring countries for two reasons. First, Thailand's land frontier is considered to have closed in the 1980s—there are simply few forests left to destroy. In the rest of the Mekong, however, the land being opened for agriculture is increasing and domestic and foreign investment in that land has increased significantly. Second, as a direct result, Thailand and others foreign investors are extending their natural resource grasp into neighboring Mekong countries, thereby externalizing the socio-ecological costs of modernization. These costs in the agro-energy context are being borne for example by Cambodia, Lao PDR, Myanmar, and Vietnam. As is amply demonstrated in the literature, the proliferation of large-scale land concessions for agricultural commodities is increasingly cause for concern.

Moreover, Thai agro-energy investment in Cambodia, Lao PDR, and Myanmar, in combination with plans to increase transportation links in the Mekong region, has already resulted in a shift in production patterns and agricultural land and water use intensity. While creating some opportunities for the local economies in neighboring countries, this research suggests that the current *business as usual* model of Thai agro-energy investment is serving to export the socio-ecological costs of production to neighboring countries.

The third objective of the research was to identify Thailand's policy options to provide alternative scenarios toward a regional interpretation of modernity in the Mekong. In this respect, four aspects are highlighted from the interviews, focus group discussions, and case narratives. First, the shift to cultivating agro-energy crops is already underway in Thailand and the Mekong region. The significant growth in agro-energy crops is illustrated in Thailand (cases 1–4), Champassak, Lao PDR (case 5), and Shan state, Myanmar (case 6). Second, assessment of the *socio-ecological* sustainability of this shift is growing and urgently required.

Without a comprehensive and evidence-based analysis of the impacts on land and water availability and use in the agricultural sector in general—not only for agro-energy crops—the Mekong region is likely to face serious social unrest and environmental consequences in the period ahead. This research indicates that sustainability must be embedded in the policy choices concerning agricultural development in general, including for agro-energy crops. This is the unequivocal message emanating from the field interviews and focus group discussions conducted during the course of this research.

Third, these developments are transpiring in the context of a global agricultural sector that is significantly distorted. These agricultural distortions persist despite ongoing multilateral negotiations in the World Trade Organization (WTO) to discipline agricultural subsidies in OECD countries and improve market access for developing countries. Critics point to the fact that the WTO has yet to make sufficient progress in removing harmful agricultural subsidies that act to both distort international trade and harm the environment. Nor has the WTO been able to secure sufficient market access for developing countries in OECD markets. Moreover, market access is further constrained in practice by technical and phytosanitary barriers. That is to say that, in principle, Thailand can export to the European market, but these exports are subject to a range of technical and phytosanitary standards and regulations. These non-tariff barriers are likely to be extended to cover the carbon footprint of traded products, further necessitating attention to the way in which products are produced, not only the characteristic of the traded products themselves. As a trading nation, Thailand's exports are impacted significantly by these measures, including based on its production and processing methods. That is one reason why the energy efficiency gains in the production process from better use of agricultural residues are an important trade issue for the Thai agricultural sector as a whole.

Fourth, the groundwork has been laid to enable greater economic integration at the Mekong regional level, with the potential to engage local level stakeholders and private sector actors in trade and investment. The portfolio of policy options outlined reflects the need to engage the policymaking process to counter the trend at all levels—local, national, regional, and multilateral to manage the natural resource base more sustainably and equitably. This is an argument for regulation, quality control, and enforcement, while simultaneously tackling lack of transparency and endemic corruption. This is why this research argues for a mechanism for policymaking to assess the overall regional landscape for agro-energy taking into account agro-ecological zones. Targeting policies to address the ecological sustainability of these zones would represent a step forward.

To date, growth and development have been solely conceived of in terms of gross domestic product. However, strategies to sustainably develop land and natural resources in the Mekong need to go beyond this classical measurement of growth, it is argued in this research, to encompass the many aspects of the overall agro-energy picture. From this perspective, the economic consequences of climate change also serve to encourage governments in the Mekong to find ways to improve resilience and decrease vulnerability in the agricultural sector. Thailand's leadership is needed to guide the Mekong region through the maze of development hurdles to better consolidate economies that are at the same time post-industrial and rural.

The Global–Local Nexus: Seeking Common Ground at the Regional Level

If past practices are deemed to be unsustainable, an alternative sustainable path forward is less clear-cut. There is, however, great promise emerging from a refocus on the agricultural sector and from rebalancing small-scale, bottom-up approaches to development. This is the lesson emerging from the case narratives elaborated in this research. Field research for this thesis in central, northeastern, and southern Thailand has illuminated a wealth of opportunities in those areas for smallholder palm oil and cassava. The narratives include farmers shifting from heavily fertilized and irrigated fruit orchards in Rangsit to less chemical and water-intensive oil palm cultivation; a community cooperative in Aoluk, Krabi being able to improve local incomes through oil palm. It reveals smallholder cassava farmers using local oil nuts to fuel local water pumps and small-scale agricultural machinery in Vanghinlad, Chumpae. These narratives illustrate that the bioenergy debate would benefit from more nuance to offer realistic opportunities to developing countries to tackle agricultural development, energy poverty, and sustainable natural resource management.

In this respect, the agricultural sector is of prime importance to generate incomes in the Mekong, while suffering from the greatest trade discrimination in the global marketplace. This is why a renewed focus on the agricultural sector is so vital to socio-ecological sustainability. This research lends weight to arguments made in the literature that global responses are not working and local solutions may be insufficient to bring about a paradigm transformation to sustainability (Halle and Raskin 2010). From this perspective, regional alternatives are emerging as a potential way to bridge the policy gap between sustainable development policy and practice.

Local narratives need to take heed of the lessons learned from agro-energy. These lessons have been clearly enunciated since the 2008 food–fuel crisis and need to be reflected in national and regional debates on the direction of agro-industry trade and investment. Notably, there has yet to be a sufficient national debate on agro-energy in Thailand and certainly not in neighboring countries. Moreover, the interviewees indicated uncertainty about agro-energy linked with policy coordination amongst the diverse ministries and levels that inform the decision-making process. This is one of the reasons why the majority (57.3%) of respondents to the survey questionnaire conducted for this research considered that their country's bioenergy policy is not heading in the right direction and that there was a lack of sustained political leadership.

There are three interlinked considerations raised by the respondents during the research. First, two-thirds of Thailand's sugar production is exported in a raw or refined state, leaving significant potential to *add value* to the supply chain through refining sugar and cassava into fuel (Fig. 3.6). Second, as the agro-energy sector grows, the share of surplus production available for fuels will also grow. Therefore, with projections for surplus ethanol production over the next several years and vague deadlines for mandated gasohol, there is a need for regulatory change to allow flexibility to export. Third, incorporation of sustainability standards to address socio-environmental aspects of bioenergy along the supply chain is facilitating greater resource efficiency. These predominantly voluntary standards and CSR are enabling a transition to a more sustainable agro-energy future in the Mekong region.

The following issues emerged from the survey undertaken for this research:

(a) Thailand has in place a visionary plan to develop renewable energy. Whether this plan succeeds in capturing the opportunities depends, to a great extent, on the institutional setting and implementation. This is why the majority of respondents (68%) in the research survey considered that the main challenges to the development of the agro-energy sector are a combination of a *consistent policy framework* and *policy implementation*, as opposed to improving crop yields, harvesting techniques or processing technologies.
(b) As a major exporter of food, Thailand's agro-energy policies have implications beyond its borders for the global supply of food. This is a complex issue that needs to be studied further.
(c) The local context matters. In the Mekong, the substantial number of small-scale farmers involved in agriculture has created a distinct narrative.[11] Agro-energy has a role to play in generating energy and employment in a region in which the vast majority are smallholders.
(d) The development of first-generation biofuels (ethanol, and biodiesel from agricultural crop biomass, such as sugar, cassava, and palm oil) can assist in the transition to a more sustainable low-carbon energy scenario if sustainability criteria are developed and implemented. To this end, South–South cooperation can stimulate knowledge building and technology transfer (e.g., with Brazil through the G20 initiative).[12]
(e) The alternative development strategy, "small is smart", put forward based on the data collected for the research survey, is to build on the synergies between small-scale initiatives and enhanced agro-energy sustainability. Small-scale community biodiesel production and use has the potential to

[11]The Mekong narrative for agro-energy is well documented in LaoFAB, an information network, moderated by Andrew Bartlett, between practitioners and academics in the region and beyond. LaoFAB is a forum for sharing information about agriculture, rural livelihoods and natural resource management in Lao PDR. The forum consists of a Google discussion group, an online library, a Facebook page, and a Linkedin group See https://www.facebook.com/LaoFAB/info/?tab=page_info.

[12]The agro-ecological (no-till, intercropping) model of Altieiri and Ernst Gotsch is bearing fruit. Fazenda da Toca in the state of São Paulo in Brazil is scaling up successful practices. See the video at https://www.youtube.com/watch?v=gSPNRu4ZPvE&feature=share. Accessed August 10, 2016.

empower smallholders with energy sufficiency, thereby lowering their energy input costs and increasing their incomes by adding value to agricultural production. This is the preferred model indicated by discussion respondents for this survey to promote bioenergy investment in the Mekong. To date, however, Thai agro-energy investment through contract farming in the region has tended to outsource environmental degradation to neighboring countries (e.g., Cambodia, Lao PDR, and Myanmar).

(f) Improving infrastructure in the Mekong region will contribute to enhancing agricultural yields and lowering production and transportation costs (e.g., water systems, energy grids, and roads). However, the results of the survey for this research indicate that policy support is required to integrate small-scale agricultural holders in the bioenergy supply chain.

(g) Governance is a vital element in addressing socio-environmental sustainability. To assist in the development of a coherent policy framework, Thailand benefits from a central body, the National Biofuels Committee, to coordinate involvement of the many ministries and private and public sector actors.[13] An institutional architecture at the Mekong regional level would facilitate integrated bioenergy development and contribute to diversifying the region's energy mix, while coordinating a regional transition toward climate-friendly agriculture and low-carbon economies.

(h) Policy space is needed to design policies that work for the region so that policymakers can assess the explicit trade-offs between water, energy, and food systems.[14] The process will be complicated increasingly by the consequences of climate change (e.g., the Mekong experienced one of the worst droughts to date in 2015).

Conclusion: Global Trends, Local Definitions

There are fish in the water and rice in the fields

Famous adage on the 13th century stele of King Ramkhamhaeng

The challenges of the twenty-first century for Southeast Asia, in particular, will be to satisfy rising demand for food and energy with less of a carbon and water

[13]During the formative years (2000–2006), the National Biofuels Committee played a crucial role in policy setting, with representation from all branches of the government, independent institutions, the private sector and academia. The Committee was abolished by the Surayuth government and the issues were delegated to several departments, with the Ministry of Energy's DEDE as a coordinating agency.

[14]To this end, since its establishment in 2011, the Greater Mekong Forum on Water, Food & Energy (2015) is exploring these linkages and trade-offs.

footprint. The quote above from King Rambkhamhaeng invoked the bounty of nature at the time of the formation of the new state of Siam in the thirteenth century. This bounty has been fundamentally altered, transforming not only Thailand, but also the Mekong from a region of resource abundance to one in which resources are "finite, threatened and fragile" (Pasuk 2000). It is also a region that is vulnerable to the impacts of climate change. Thailand ranked 9 out of 187 countries most affected by extreme weather events between 1995–2014, while Vietnam ranked 7, and Myanmar ranked 10 (Global Climate Risk Index 2016).

In the face of the urgency of climate change, there is evidence that small-scale bioenergy in the Mekong has the potential to shift the agricultural sector away from high-input, energy-intensive agriculture towards more sustainable practices. In order to do so, smallholders require an enabling policy architecture to empower local communities to generate their own energy for consumption both on and off the national grid. In this respect, the Clean Development Mechanism (CDM) has added economic viability and environmental motivation to bioenergy initiatives. It is stimulating investment and innovation in energy from first-generation sources (crop residues and biogas) and second-generation cellulosic biofuels.

Over the coming decades, the world economy will change radically as markets move to reflect scarcities in food and fuel. As a major contributor to the global supply of agricultural products, Thailand faces a complex range of factors in implementing its vision for a low-carbon economy. Thailand has the potential to move beyond past inefficiencies and take up a leadership role in developing renewable energy in the region. If the current stimulus to produce and use first-generation biofuels brings about a transition to more technologically complex second-generation systems (e.g., bagasse, algae), there is likely to be less competition between fuel and food production.

Given that demand for bioenergy is driven largely by regulatory mandates, with production costs subsidized by governments, there are valid issues raised relating to their economic efficiency and socio-environmental sustainability. Concerns related to the commercialization of biofuels alongside small-scale initiatives warrant more deliberate attention, particularly at the regional level. Evidence from the fieldwork conducted for this research suggests that the model of developing energy from biomass at the community level may be a sufficiently solid basis to allow Thailand to take the lead in orchestrating a regional shift to low-carbon economies.

However, we find gaps between the mounting evidence of bioenergy practices and policies for the dual track development of community and commercial bioenergy systems. In other words, whilst the development of bioenergy may be beneficial for Thailand, the implications for sustainable agricultural practices in neighboring Mekong countries are not necessarily as favorable. The challenge is to bridge this gap to enable the transition toward a more sustainable energy future for the Mekong region.

Thailand's experience with agro-energy may serve, in turn, to assist neighboring countries in the Mekong region to enhance their renewable energy development options. Whilst national debate invariably ends on an optimistic note concerning the prospects for renewable energy in general, it is worthwhile noting that there has yet to be a sufficiently rigorous debate in the Mekong on the merits of agro-energy. This may explain why the majority of respondents (57.3%) in the survey undertaken for this research considered that the bioenergy sector is *not* developing along the right path. To this end, all the 143 respondents surveyed and many interviewees felt the government could benefit from more information and improved coordination between ministries in formulating biofuels policies. This is particularly the case given the need for complex trade-offs needed to formulate sustainable agro-energy policies. Moreover, our research revealed that consideration of climate change as a driver for changes in policies and practices was lacking. It is crucial to verify if this continues to be the case.

This chapter has argued that there is policy space to define local narratives in response to global problems. By the same token, it bears emphasizing that these narratives need not repeat mistakes learned in other countries, for example, with respect to subsidizing biofuels, nor need they ignore conventional wisdom. The experience of the United States has been subject to criticism for the abundant use of subsidies given to the corn ethanol industry (Koplow 2007), as has the European Union for subsidies to rapeseed (Jung et al. 2010). It has been well-documented that these subsidies have distorted markets and increased production (IISD/GSI 2015; Steenblik 2007) to the detriment of the environment and led to an extensive application and over-use of pesticides, contaminating water supplies and depleting the soil of nutrients over time (IISD/GSI 2012; Searchinger 2009).

This leaves us with a fundamental question underlying this research relating to the policy space for countries to determine policies. To what extent is Thailand positioned to reap the potential benefits of developing a viable bioenergy sector; or, will it repeat the mistakes of those countries that have put in place burdensome subsidies and allowed agro-industrial production to crowd out smallholder agricultural production? Further evidence is needed to determine whether the future agro-energy narrative for the Mekong can be built on locally-driven initiatives such as the ones studied in this research. Beyond the Mekong, many developing countries in Africa and Latin America are also exploring the potential synergies from linking climate mitigation and adaptation with innovative agro-energy initiatives.

Significant improvements in policy and practice are required to reconfigure agricultural production and land use in order to meet the global demand for food and fuel in a way that contributes to food security, energy generation, greenhouse gas emissions reduction, and biodiversity conservation. In this regard, we would like to underscore four important points that arise out of the above analysis in relation to the context for bioenergy development. The first involves the scale of biofuels development. The second involves the objectives underlying biofuel development. The third is linked with the actors involved in biofuel production relating to the distinction between commercial and community biofuels. The fourth and, arguably the most important element, concerns policy space and governance capacity. Effective governance of the policy space, in turn, will impact the other elements; Simply put, how much, for whom, by whom, and how.

These four caveats are fundamentally important when considering the way forward for agro-energy in the Mekong. The regional energy narrative is currently being formulated, with trade in bioenergy an option under consideration. The driving factors behind the current development of bioenergy are open to narrower and broader interpretations depending upon the context in which they are advocated. While it is necessary to improve the economics and monitor environmental sustainability, the argument of bioenergy proponents in the Mekong and other developing countries, notably Brazil, is that the basket of benefits outweighs the constraints—at least in the transition to a low-carbon economy.

References

Abdel Motaal, D. (2008). The biofuels landscape: Is there a role for the WTO? *Journal of World Trade, 42*(1), 61–86.

Achawangkul, Y. (2015). *Thailand's integrated energy blueprint*. Bangkok: Department of Alternative Energy Development and Efficiency.

Asian Development Bank (ADB). (2009a). *Road map for expanded energy cooperation in the Greater Mekong*. Manila: ADB.

ADB. (2009b). *Development partners meeting on GMS biofuels and rural renewable energy*. Bangkok: ADB.

ADB. (2009c). *Status and potenial for the development of biofuels and rural renewable energy: Thailand*. Manila: ADB.

ADB. (2009d). *Exploring core environment program-private sector partnerships for developing biodiesel as an alternative fuel in the Greater Mekong Subregion*. Bangkok: ADB.

ADB. (2013). *Thinking about water differently: Managing the water-food-energy nexus*. Manila: ADB.

ADB. (2015). *Fossil fuel subsidies in Thailand: Trends, impacts and reform*. Manila: ADB. Retrieved August 10, 2016, from http://www.adb.org/sites/default/files/publication/175455/fossil-fuel-subsidies-thailand.pdf.

Allouche, J., Middleton, C., & Gyawali, D. (2015). Technical veil, hidden politics: Interrogating the power linkages behind the nexus. *Water Alternatives, 8*(1), 610–626.

Altieiri, M. (2009). Agroecology, small farms, and food sovereignty. *Monthly Review*, July–August.

Apichart, J. (2015). Integrated energy development. Presentation for the 12th Annual Meeting of the Working Group on Agriculture (WGA). Bangkok: WGA.

Babel, M. S., Shrestha, B., & Perret, S. R. (2011). Hydrological impact of biofuel production on hydrology: A case study of the Khlong Phlo Watershed in Thailand. *Agricultural Water Management, 101*, 8–26.

Beck, U. (1992). *Risk Society: Towards a new modernity*. Translation by M. Ritter. London: Sage. Originally published in 1986 as *Risikogesellschaft: Auf dem Weg in eine andere Moderne*. Frankfurt a. M.: Suhrkamp.

Brazil. (1998). The energy sector: The case of alcohol fuel (Ethanol). *Submission by Brazil to the World Trade Organisation (WTO) Committee on Trade and Environment*. Geneva: WTO.

Brazil. (2005). Environmental goods for development. *Submission by Brazil to the WTO (TN/TE/W/59)*. Geneva: WTO.

Bruckman, V., Haruthaithanasan, M., Kraxner, F., Miller, F., Darabant, A., Choumnit, G., et al. (2016). *ACMECS Bioenergy 2015: Three years of efforts towards a regional bioenergy network*. Bangkok: International Union of Forest Resource Organizations.

Bundit, F. (2009). Recent policies and challenges in sustainable energy development in Thailand. In V. K. Vijay & H. P. Garg (Eds.), *Renewable energy and environment for sustainable development* (pp. 12–25). New Delhi: Narosa.

Chantanakome, W. (2016). *Regional energy security infrastructure Perspective: Strategic policy and action*. Bangkok: Presentation at sustainable energy and technology Asia (SETA). http://www.eppo.go.th/index.php/th/planpolicy/tieb/aedp

Chanthawong, A., & Dhakal, S. (2016). Liquid biofuels development in Southeast Asian countries: An analysis of market, policies and challenges. *Waste and Biomass Valorization, 7,* 157.

Charnovitz, S., Earley, J., & Howse, R. (2008). *An examination of social standards in biofuels sustainability criteria*. Washington D. C: International Food and Agricultural Trade Policy Council.

Cramer Commission. (2007). *Testing framework for sustainable biomass*. Utrecht, The Netherlands: Cramer Commission.

DEDE. (2013). *Alternative energy development plan*. Bangkok: Ministry of Energy. Retrieved August 10, 2016, from http://www.dede.go.th/download/state_58/sit_57_58/Thailand-Alternative-Energy-Situation.pdf.

Department of alternative energy development and efficiency (DEDE). (2015a). *Ten-year renewable energy development plan (2015–2025)*. Bankgok: Ministry of Energy.

DEDE. (2015b). *Energy in Thailand: Facts and figures for 2014*. Bangkok: Ministry of Energy. Retrieved August 10, 2016, from http://weben.dede.go.th/webmax/sites/default/files/factsq1_2014.pdf.

Dehue, B., Meyer, S., & Hamelick, C. (2007). *WWF—Towards A harmonised sustainable biomass certification*. Utrech, The Netherlands: WWF.

Doornbosch, R., & Steenblik, R. (2007). *Biofuels: Is the cure worse than the disease?*. Paris: Organisation for Economic Cooperation and Development.

EC. (2009). Directive 2009/28/EC on the promotion of the use of energy from renewable sources (repeal. Directives 2001/77/EC and 2003/30/EC. Directive 2008/0016). 23 April. *Official Journal of the European Union*, 5 June.

EC, European Commission. (2015). Directive 2015/1513 to reduce indirect land use change for biofuels and bioliquids (Amending Directive 98/70/EC relating to the quality of petrol and diesel fuels and 2009/28/EC on the promotion of the use of energy from renewable sources). 9 September. *Official Journal of the European Union*, 15 September.

Echols, M. (2009). *Biofuels certification and the law of the WTO*. Geneva: International Center for Trade and Sustainable Development.

ECOFYS. (2009). *Trade sustainability impact assessment for the free trade area between the European Union and the Association for Southeast Asian Nations*. Rotterdam: ECOFYS.

ECOFYS. (2016). *Greenhouse gas reduction roadmap for Thailand*. Rotterdam: ECOFYS and Global Green Growth Institute.

Food and Agriculture Organization (FAO). (2008). *State of food and agriculture. Biofuels: prospects, risks and opportunities*. Rome: FAO. http://www.ftp.fao.org/docrep/fao/011/i0100e/i0100e.pdf.

FAO. (2009a). *Enabling agriculture to contribute to climate change mitigation*. Submission to the Ad Hoc Working Group on Long-term Cooperative Action under the Climate Change Convention. Rome: FAO. Retrieved August 10, 2016, from http://www.unfccc.int/resource/docs/2008/smsn/igo/036.pdf.

FAO. (2009b). *Small-scale bioenergy initiatives*. Rome: FAO & Policy Innovation Systems for Clean Energy Security.

FAO. (2011). *Launch of the state of the world's land and water resources for Food and Agriculture (SOLAW): Managing systems at risk*. Rome: FAO. Retrieved August 10, 2016, from http://www.fao.org/nr/water/news/solaw_launch.html.

FAO. (2012a). *Policies and institutions to support smallholder agriculture*. Rome: FAO.

FAO. (2012b). *Smallholders in global bioenergy value chains and certification*. Rome: FAO. http://www.fao.org/3/a-i2597e.pdf.

Foran, T. (2013). Impacts of natural resource-led development on the Mekong Energy System. In A. Smajgl & J. Ward (Eds.), *The water-food-energy nexus in the Mekong Region* (pp. 105–142). New York: Springer.

Fullbrook, D. (2007). *Contract farming in Lao PDR: Cases and questions*. Vientiane: Laos Extension for Agriculture Project (LEAP).

Fullbrook, D. (2013). Food security in the wider Mekong Region. In A. Smajgl & J. Ward (Eds.), *The water-food-energy nexus in the Mekong Region* (pp. 61–104). New York: Springer.

Giampietro, M., & Mayumi, K. (2009). *The biofuel delusion: The fallacy of large scale agro-biofuels production*. London: Earthscan.

Global Carbon Atlas. (2017). Retrieved April 5, 2018, from http://www.globalcarbonatlas.org/en/CO2-emissions.

Global Climate Risk Index (GCRI). (2016). Bonn: Germanwatch. Retrieved August 10, 2016, from https://germanwatch.org/fr/download/13503.pdf.

Gotsch, E. (2016). Fazenda da toca, state of São Paulo. https://youtu.be/gSPNRu4ZPvE.

Greater Mekong Forum on Water, Food & Energy. (2015). *Proceedings*. Implemented by the CGIAR Resource Programme on Water, Land and Ecosystems. https://wle-mekong.cgiar.org/2015forum/.

Guariguata, M., Masera, O., Johnson, F. X., von Maltitz, G., Bird, N., Tella, P., et al. (2011). *A Review of environmental issues in the context of biofuels sustainability*. Bogor: Center for International Forestry Research.

Halle, M., & Raskin, P. (2010). *World trade: A new direction*. The global trade inititative. Boston: Tellus Institute. http://www.tellus.org/pub/GTI%20Perspective%20-%20World%20Trade%20-%20A%20New%20Direction.pdf.

Hodbodt, J., & Tomei, J. (2013). Demystifying the social impacts of biofuels at local levels: Where is the evidence? *Geography Compass*, 15 July.

Huang, J., Qiu, H., Yang, J., & Zhang, Y. (2009). *Status and potential for the development of biofuels and rural renewable energy: The People's Republic of China*. Manila: ADB.

IAASTD, International assessment of agricultural knowledge, science and technology for development. (2009). *Agriculture at a crossroads: Summary for decision makers of the global report*. Report for the United Nations. Washington, D.C.: Island Press.

IEA. (2011). *Technology roadmap: Biofuels for transport*. Paris: IEA. Retrieved August 10, 2016, from http://www.iea.org/publications/freepublications/publication/Biofuels_Roadmap_WEB.pdf.

IEA. International Energy Agency. (2015). *World energy outlook 2015*. Paris: IEA.

IISD/GSI. (2012). *State of play on biofuels subsidies: Are policies ready to shift?*. Geneva: IISD/GSI.

IISD/GSI. (2015). *Financing the sustainable development goals through fossil-fuel subsidy reform: Opportunities in Southeast Asia, India and China*. Winnipeg: IISD. Retrieved August 10, 2016, from http://www.iisd.org/gsi/sites/default/les/nancing-sdgs-fossil-fuel-subsidy-reform-southeast-asian-india-china(6).pdf.

Jitsanguan, T. (2001). *Sustainable agricultural systems for small-scale farmers in Thailand: Implications for the environment*. Bangkok: Department of Agirucltural and Resource Economics, Kasetsart University. http://www.fftc.agnet.org/htmlarea_file/library/20110718190247/eb509.pdf.

Johnson, F. X., & Seebaluck, V. (Eds.). (2012). *Bioenergy for sustainable development and international competitiveness*. New York: Routledge.

Jung, A., Dörrenberg, P., Rauch, A., & Thöne, M. (2010). *Biofuels—At what cost? Government support for ethanol and biodiesel in the European Union–2010 Update*. Geneva: IISD/GSI.

Koplow, D. 2007. *Biofuels–at what cost? Government support for ethanol and biodiesel in the United States–2007 Update*. Geneva: IISD/GSI. http://www.iisd.org/gsi/biofuel-subsidies/biofuels-what-cost.

Kumar, S., Salam, P. A., Shrestha, P., & Ackom, E. (2013). An assessment of Thailand's biofuel development. *Sustainability, 5*, 1577–1597.

Lopez, G. P., & Laan, T. (2008). *Biofuels—At what Cost? Government support for ethanol and biodiesel in China*. Geneva: IISD/GSI.

Malik, U., Ahmed, M., Sombilla, M.A., & Cueno, S. (2009). Biofuels production for smallholder producers in the Greater Mekong Sub-region. *Applied Energy, 86,* 58–68

Nguyen, T., Gheewala, S., & Garivait, S. (2008). Full chain energy analysis of fuel ethanol from cane molasses in Thailand. *Applied Energy, 85,* 722–734.

Opijnen, M., van & Oldenziel, J. (2011). *Responsible supply chain management.* Report prepared for the European Commission High Level Group on Corporate Social Responsibility. Brussels: Centre for Responsible Multinational Corporations.

Pasuk, P. (2000). State decisions and community rights. In A. Nozakik & C. Baker (Eds.), *Village communities, states, and traders: Essays in honour of Chatthip Nartsupha* (pp. 188–194). Bangkok: Sangsan.

Penunia, M. E. (2009). Upscale sustainable, integrated, diversified, organic agriculture by smallholders farmers to adapt and mitigate effects of climate change. Paper presented by the Asian Farmers' Association for sutainable rural development at the 32nd session of IFAD's Governing Council.

Pimental, D. (2006). Soil erosion: A food and environmental threat. *Environment, development and sustainability,* 119–137.

Pimental, D. (2009). Food versus biofuels: Environmental and economic costs. *Human Ecology, 37*(1), 1–12.

Potts J., Lynch M., Wilkings A., Huppé G.A., Cunningham M., & Voora V. (2014). *The state of sustainability initiatives review 2014: Standards and the green economy.* Geneva: IISD and IIED. Retrived August 10, 2016, from http://www.iisd.org/library/state-sustainability-initiatives-review-2014-standards-and-green-economy.

Roundtable on Sustainable Biomaterials (RSB). (2011). *Consolidated RSB EU RED principles & criteria for sustainable biofuel production* (Version 2.1). http://rsb.org/pdfs/standards/RSB-EU-RED-Standards/13-03-01-RSB-STD-11-001-01-001%20vers%202.1%20Consolidated %20RSB%20EU%20RED%20PCs.pdf.

Roundtable on Sustainable Palm Oil (RSPO). (2016). Impacts. http://www.rspo.org/about/impacts.

Salvatore, M., & Damen, B. (2010). *Analysis for Thailand. Bio energy and food security project.* Rome: FAO.

Samai, J.-I. (2016). *Vision and perspectives of the biofuel sector in Thailand.* Bangkok: Presentation at Sustainable Energy and Technologia Asia (SETA), March.

Searchinger, T. (2009). Government policies and drivers of world biofuels, sustainability criteria, certification proposals and their limitations. In R. Howard & S. Bringezu (Eds.), *Biofuels: Environmental consequences and interactions with changing Land use* (pp. 37–52). Ithaca: Cornell University Press.

Segschneider, K. (Ed.). (2008). *Field survey on sustainable palm oil in Thailand and compliance with international standards.* Bangkok: Southeast Asia Consult & Resource Company.

SEI, Stockholm Environment Institute. (2008). *Bioenergy for sustainable development and global competitiveness: The case of sugar cane.* Stockholm: SEI.

Shaw, S. (2009). Bioenergy policies and legal framework in Thailand. In E. Morgera, K. Kulovesi, & A. Gobena (Eds.), *Case studies on bioenergy policies and law: Options for sustainability* (pp. 289–328) Rome: FAO. http://www.fao.org/docrep/012/i1285e/i1285e.pdf.

Shaw, S. (2010). *Trade in agro-energy in the Mekong.* Bangkok: Chulalongkorn.

Silalertruksa, T., & Gheewala, S. (2012). Environmental sustainability assessment of palm biodiesel production in Thailand. *Energy, 43,* 306–314.

Silalertruksa, T., Gheewala, S., Hunecke, K., & Fritsche, U. R. (2012a). Biofuels and employment effects: Implications for socioeconomic development in Thailand. *Biomass and Bioenergy, 46,* 409–418.

Silalertruksa, T., Bonnet, S., & Gheewala, S. (2012b). Life cycle costing and externalities of palm oil biodiesel in Thailand. *Journal of Clean Production, 28,* 225–232.

Siriwardhana, M., Opathella, G., & Jha, M. (2009). Biodiesel: Initiatives, potential and prospects in Thailand. *A Review Energy Policy, 37,* 554–559.

Smajgl, A., & Ward, J. (Eds.). (2013). *The water-food-energy nexus in the Mekong Region.* New York: Springer.

Sombilla, M., Malik, U., Ahmed, M., & Cueno, S. (2009). *Integrating biofuel and rural renewable energy production in agriculture for poverty reduction in the Greater Mekong Subregion*. Manila: ADB.

Steenblik, R. (2007). *Biofuels–at what cost? Government support for ethanol and biodiesel in selected OECD countries*. Geneva: IISD/GSI. Retrieved from http://www.iisd.org/gsi/biofuel-subsidies/biofuels-what-cost

Surapong, C. (2008). *Fuel ethanol market assessment and potential of cassava in Thailand*. Bangkok: Field Crop Research Institute, Department of Agriculture.

TGO, Thailand Greenhous Gas Organisation. (2016). TGO carbon monthly. www.tgo.org.

Thamsiriroj, T., & Murphy, J. (2009). Is it better to import palm oil from Thailand to produce biodiesel in ireland than to produce biodiesel from indigenous Irish rape seed? *Applied Energy, 86,* 595–604.

Tharakan, P., Crishna, N., Romero, J., & Morgado, D. (2012). *Biofuels in the greater Mekong subregion: Energy sufficiency, food security, and environmental management*. ADB Southeast Asia Working Paper Series No. 8. Manila: ADB.

Tilman, D., Socolow, R., Foley, J. A., Hill, J., Larson, E., Lynd, L., et al. (2009). Beneficial biofuels: The food, energy, and environment trilemca. *Science, 325,* 270–271.

UNCTAD, United Nations Conference on Trade and Development. (2008). *Biofuel production technologies: Status, prospects and implications for trade and development*. New York: UNCTAD.

UNE, United Nations-Energy. (2007). *Sustainable bioenergy: A framework for decision makers*. New York: UNE.

UNEP, United Nations Environment Programme. (2011). *The bioenergy and water nexus*. Nairobi: UNEP Bioenergy, Oeko Institute and IEA Bioenergy. http://www.unep.org/bioenergy/Portals/48107/publications/waternexus_full.pdf.

UNESCAP, United Nations Economic and Social Commission for Asia and the Pacific. (2009). *Sustainable agriculture and food security in the Asia and Pacific*. Bangkok: UNESCAP.

Uriarte, F. (2010). *Biofuels from plant oils*. Jakarta: ASEAN Foundation. Retrieved August 10, 2016, from http://www.aseanfoundation.org/documents/books/biofuel.pdf.

USAID, United States Development Agency. (2009). *Biofuels in Asia: An analysis of sustainability options*. Bangkok: USAID.

USDA, United States Department of Agriculture. (2016). *Thailand biofuels annual*. Bangkok: USDA Foreign Agricultural Service. http://gain.fas.usda.gov/Recent%20GAIN%20Publications/Biofuels%20Annual_Bangkok_Thailand_7-22-2016.pdf.

Weerapat Sessomboon. (2014). *Map of clustering of ethanol distribution in Thailand*. Correspondence with author.

World Bank. (2008). *World development report 2009: Agriculture for development*. Washington, D.C.: World Bank.

World Bank. (2016). *Thailand economic monitor*. Bangkok: World Bank.

WWF, World Wide Fund for Nature. (2016). Power Sector Vision 2015. http://d2ouvy59p0dg6k.cloudfront.net/downloads/wwf_greater_mekong_power_sector_vision_2050.pdf.

WWF & SNV, The Netherlands Development Organisation. (2009). *Developing sustainable pro-Poor biofuels in the Mekong Region*. Vientiane: WWF and SNV.

Yang, J., Huang, J., Qiu, H., Rozelle, S., & Sombilla, M. (2009). Biofuels and the Greater Mekong Subregion: Assessing the impacts on prices, production and trade. *Applied Energy, 86,* 537–546.

Zarrilli, S., & Burnett, J. (2008). *Making certification work for sustainable development: The case of biofuels*. Geneva: UN Conference on Trade and Development. Retrieved August 10, 2016, from http://www.irgc.org/IMG/pdf/Wirec-Simonetta_Zarrilli.

Part II
Eco-resources and Their Economic Significance

Chapter 4
Toward Carbon Certificate in Vietnam: Net Ecosystem Production and Basic Income for the Local Community

Tran Van Do and Tamotsu Sato

Abstract In recent years, ecologists have focused on estimating Net Ecosystem Production (NEP) to understand the role of forests against increasing concentrations of CO_2 in the atmosphere, a major concern in researches and debates on global warming. This is due to the fact that NEP of a forest is the amount of carbon accumulated in a unit of area and time. This chapter discusses NEP in tropical broad-leaved forests of the Copia Natural Reserve, northwest Vietnam and the need for a carbon certificate to support the local community. Based on field research in northwest Vietnam, this chapter proposes a modified and easily applicable method for estimating NEP. Research results indicate that one hectare of secondary broad-leaved forest in Vietnam can accumulate 6.57 Mg C y^{-1}, more than twice that of old-growth forest at 2.57 Mg C y^{-1}. Research suggests that NEP is higher than some other forests around the world. We suggest that a price should minimally sustain the livelihood of forest protectors, at around 10 US$/ton carbon, a price much higher than current government payments of 10 US$ ha y^{-1}, regardless of carbon accumulation. This chapter asserts the importance of issuing forest carbon certificate to forest protectors so that local communities can raise their bargaining power to improve their income while at the same time, protect the natural forests sustainably.

Keywords Carbon market · Carbon sequestration · Poverty reduction
Southeast Asia · Tropical forests

T. Van Do (✉)
Research Institute for Sustainable Humanosphere (RISH), Kyoto University, Kyoto, Japan
e-mail: dotranvan@hotmail.com

T. Van Do
Silviculture Research Institute, Vietnamese Academy of Forest Sciences, Hanoi, Vietnam

T. Sato
Department of Forest Vegetation, Forestry and Forest Products Research Institute,
Tsukuba, Japan

M. Lopez and J. Suryomenggolo (eds.), *Environmental Resources Use and Challenges in Contemporary Southeast Asia*, Asia in Transition 7,
https://doi.org/10.1007/978-981-10-8881-0_4

Introduction

Net Ecosystem Production (NEP) is a fundamental property of ecosystems. It was originally defined by Whittaker and Woodwell (1968) as the difference between the amount of organic carbon fixed by photosynthesis in an ecosystem (gross primary production) and total ecosystem respiration, Re (the sum of autotrophic and heterotrophic respiration). Based on this definition, NEP represents the organic carbon available for storage within the system or loss from it by export or nonbiological oxidation. In other ways, NEP is usually described as the balance between Net Primary Production (NPP) and heterotrophic respiration in an ecosystem. Therefore, NEP is known as the rate of carbon accumulation in forest ecosystem (Caspersen et al. 2000; Randerson et al. 2002).

This chapter proposes a modified and easily applicable method for estimating NEP in forests[1]. Conventionally, there are four main components for estimating NEP (see Fig. 4.1), i.e., annual living biomass increment, annual dead biomass storage, biomass consumed by herbivores, and heterotrophic respiration[2]. All of these four components were measured and calculated over one year of fieldwork (2014–2015) on site in Vietnam, as described in the appendix below. Our method to collect all the data that is parallel to other researches conducted elsewhere, with a little adaption to local conditions. All data analyses were conducted by using SAS 9.2 (SAS Institute Inc., Cary, NC, USA). This chapter starts with a description of the research site and a discussion of our research methodology. This is then followed by a discussion of the domestic situation of carbon pricing in Vietnam and the applicability of our estimations.

Study Site Description

Study was conducted in a tropical evergreen broad-leaved forests of Copia Natural Reserve, northwest Vietnam, at 21°23'N and 103°38'E (see Fig. 4.2). The Copia Natural Reserve covers an area of 19,745 ha, including 13,426 ha of natural forests.

---

[1]This method also has applicability to boreal, montane, and coastal forests.

[2]Annual living biomass increment includes both belowground and aboveground biomass. Identifying living biomass increment of aboveground compartment has been widely carried out. Most researches are based on allometric models between diameter at breast height (DBH) and stem biomass (Sherman et al. 2003; Fukushima et al. 2008 and many others), then the difference of biomass between times t_i and t_j is aboveground living biomass increment. Biomass increment of coarse roots (roots with diameter (φ) > 2 mm) is also estimated in the same manner. Annual dead biomass storage contains those in soil and those on forest floor. For those on forest floor, estimating method is quite simple by using litter traps distributed under forest canopy (Sato et al. 2010 and many others). For those in soil, estimating method is rather complicated. Recently, ecologists divide roots into two types, those $\varphi \leq 2$ mm, functioning as absorbing water and nutrient for trees called fine roots, and those with $\varphi > 2$ mm, functioning as stabilizing trees, called coarse roots. Dead biomass storage of fine roots is estimated by continuous inflow method (Osawa and Aizawa 2012). While, dead biomass of coarse roots is quite small, and it is usually ignored in estimating NEP. In addition, C_h is also ignored because it is known as negligible and rather difficult in estimation (Clark et al. 2001).

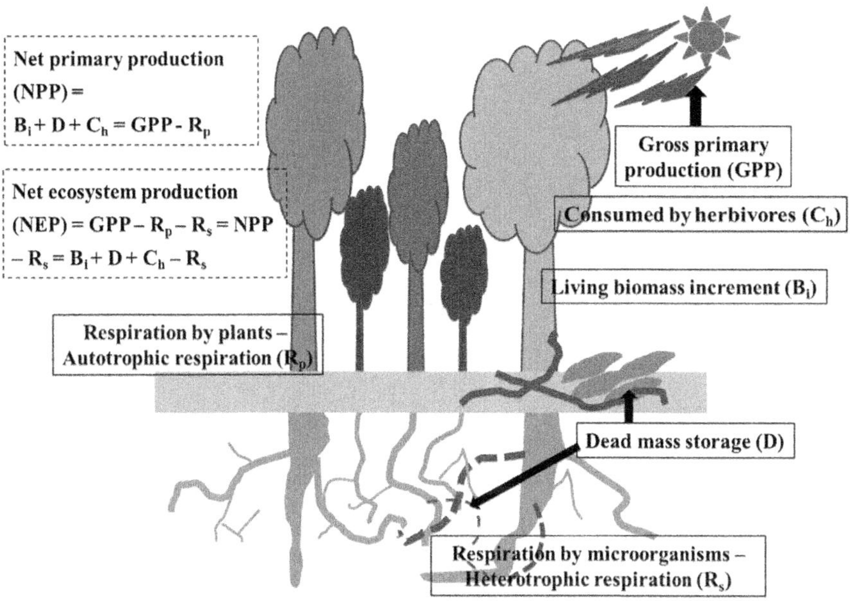

Fig. 4.1 The summation method used to estimate NEP of forests

There are two types of forests classified as old-growth forest, which have no signs of human disturbance such as logged stumps or burnt clearings, and secondary forest recovered after abandoned shifting cultivation (Tran et al. 2010, 2011). The Copia Natural Reserve is located in three communes (Co Ma, Long He, and Chieng Bom), Thuan Chau District, Son La Province. It is about 360 km far from Hanoi, capital of Vietnam. It takes about 8 hours by bus from Hanoi to the Copia Natural Reserve. In the research area, mean annual rainfall was 1277 mm and mainly fell in the summer season between May and July. Mean monthly temperature ranged from 21 to 23°C in summer, and from 12 to 16°C in winter. The annual relative humidity is 80% (Tran et al. 2005). In this area, the dominant soil type is Ferralic Acrilsols, which is acidic with low base saturation and poor in nutrient content. Humic Acrisols and Rhodic Ferralsols are also found occasionally (Nguyen 1996).

On the site, we established a 30 m × 30 m plot in old-growth forest and another plot of the same size in a 35-year-old secondary forest for NEP estimation. The age of secondary forest was estimated by coring three largest stems in the plot for identifying annual rings. The NEP or rate of carbon accumulation in a forest ecosystem is estimated based on the calculation of Net Primary Production (NPP) minus the heterotrophic respiration (soil respiration) of the forest (O'Connell et al. 2003).

In the following sections, we present the results of NPP and soil respiration and finally, the NEP estimation. Based on this NEP estimation, we would like to discuss implications for the carbon market and price, and recommend its applicability in

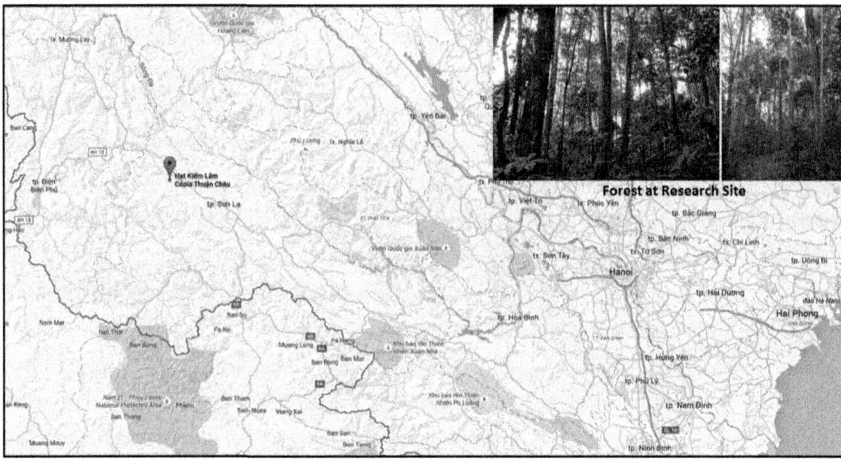

Fig. 4.2 Map of the study site

Vietnam to improve the income for local people/forest protectors. This is a crucial issue for sustaining local livelihood. It is well accepted that additional information on carbon sinks and carbon sequestration/NEP in forests worldwide is required to support REDD (Reducing Emissions from Deforestation and Forest Degradation). Earlier research has shown how REDD can be economically viable and will be accepted by providers only if payments for avoiding deforestation and degradation is at least as large as opportunity cost plus transaction costs (Wunder et al. 2008). Therefore, one of the first steps toward the implementation of REDD at a country level, in this case Vietnam, is to estimate forest carbon stocks and NEP. NEP, as estimated in this study, is crucial for REDD monitoring and subsequently, for fundraising from public funds or carbon markets. It is hoped that these results will contribute to the improvement of local people's livelihood and sustainable forest management in Vietnam.

Net Primary Production

On the site, forty species were found in a 900 m^{-2} plot in old-growth forest, while there were only 28 species in secondary forest (Table 4.1). A density of 88 and 96 stems in 900 m^{-2} was found in old-growth forest and secondary forest, respectively. Mean (Diameter at Breast Height) DBH and basal area were significantly different between the two forest stands. In addition, a largest stem in old-growth forest was 52.5 cm in DBH, which was much larger than that in secondary forest (30.9 cm in DBH).

Litterfall in the old-growth forest was higher than that in secondary forest, regardless of seasonality (see Fig. 4.3a). Meanwhile, in either old-growth forest or secondary forest, litterfall was highest in December–April, reducing in March–July, and in July–December (see Fig. 4.3a). Annual mean litterfall in old-growth forest was 4.02 g m^{-2} d^{-1} and it was 2.16 g m^{-2} d^{-1} in secondary forest (see Fig. 4.3b).

Table 4.1 Stand parameter of old-growth forest and secondary forests

Stand parameter	Old-growth forest	Secondary forest
Density (stems 900 m^{-2})	88 ± 5	96 ± 6
Mean DBH (cm)	15.5 ± 1.3[a]	13.0 ± 0.7[b]
Maximum DBH (cm)	52.5	30.9
Basal area (m^{-2} 900 m^{-2})	2.64 ± 0.5[a]	1.63 ± 0.3[b]
Species number	40	28
NPP (Mg C ha^{-1} y^{-1})	15.71	13.25
Soil respiration (Mg C ha^{-1} y^{-1})	13.14	6.68
NEP (Mg C ha^{-1} y^{-1})	2.57	6.57

[a, b]Different letters in a line indicate significant difference by *t*-test

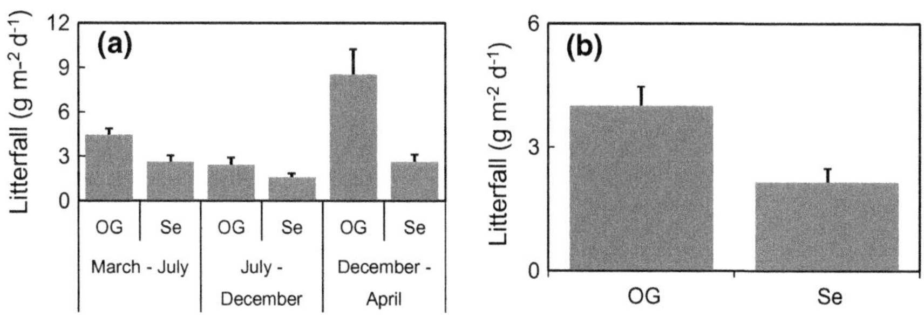

Fig. 4.3 Seasonality litterfall (**a**) and annual mean litterfall (**b**). OG is old-growth forest, Se is secondary forest

Most fine roots were distributed in 0–20 cm soil surface in both old-growth forest (see Fig. 4.4a) and secondary forest (see Fig. 4.4b); the deeper the soil, the less distribution of fine roots. In old-growth forest, 65% of fine roots were distributed in 0–20 cm soil surface, while this was 76.6% in secondary forest.

Fine root decomposition, mortality, and production were much higher in the old-growth forest (see Fig. 4.5a) compared to that in secondary forest (see Fig. 4.5b). Decomposition was 0.03 g m^{-2} d^{-1} in old-growth forest and 0.11 g m^{-2} d^{-1} in secondary forest. Mortality was 0.07 g m^{-2} d^{-1} and 0.33 g m^{-2} d^{-1} in old-growth forest and secondary forest, respectively. Meanwhile, fine root production was 1.00 g m^{-2} d^{-1} in old-growth forest, nearly three times that found in secondary forest (0.36 g m^{-2} d^{-1}).

In both forest stands, the smaller DBH classes contributed to less aboveground biomass (AGB) but higher AGB increment (see Fig. 4.6a, b). In the old-growth forest, all stems with DBH $\leq$ 40 cm contributed 44% AGB, but 58.9% AGB increment (see Fig. 4.6a). Meanwhile, there was no stem with DBH > 40 cm appearing in secondary forest, where all stems with DBH $\leq$ 20 cm contributed 35% AGB, but 56.1% AGB increment (see Fig. 4.6b). Generally, smaller stems play a more important role in AGB accumulation than larger ones.

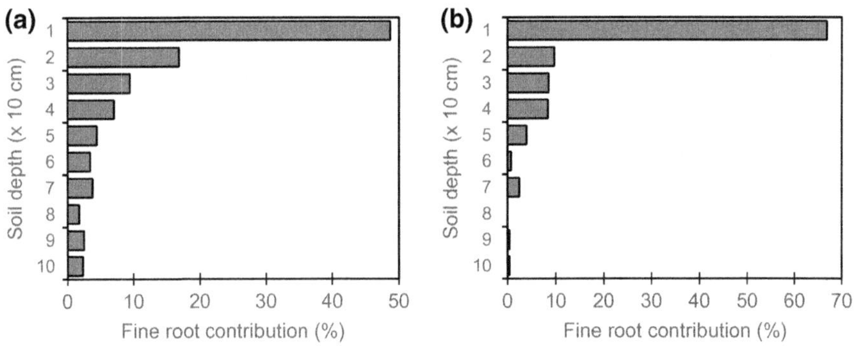

Fig. 4.4 Vertical distribution of fine roots in old-growth forest (**a**) and secondary forest (**b**)

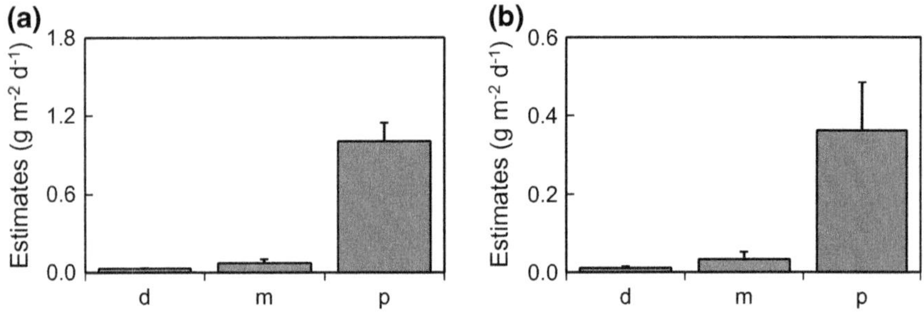

Fig. 4.5 Fine root decomposition, mortality, and production in old-growth forest (**a**) and secondary forest (**b**). d is decomposition, m is mortality, and p is production

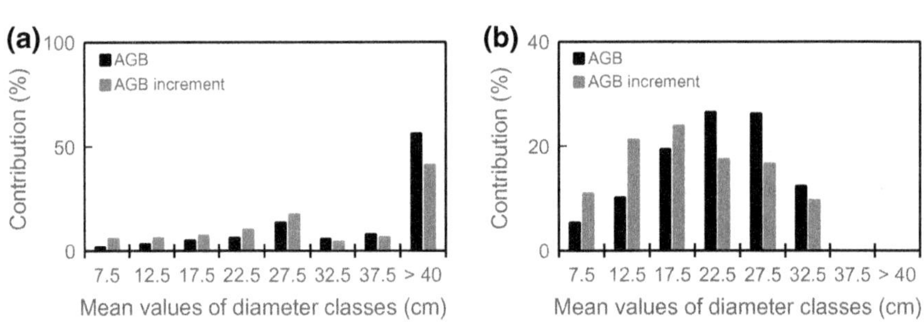

Fig. 4.6 Contribution of aboveground biomass (AGB) and AGB increment of each diameter class in old-growth forest (**a**) and secondary forest (**b**)

AGB increment in old-growth forest was 3.19 g m^{-2} day^{-1} (Table 4.2); lower than that in secondary forest of 4.13 g m^{-2} day^{-1}. The difference of AGB increment led to a difference of coarse root increment as 0.40 g m^{-2} day^{-1} in old-growth forest and 0.61 g m^{-2} day^{-1} in secondary forest.

Table 4.2 Contribution of each compartment to total NPP

	Belowground NPP		Aboveground NPP		Total
	Coarse root increment	Fine root production	Aboveground litterfall	Aboveground stand increment	
Old-growth forest					
g biomass m^{-2} day^{-1}	0.40 ± 0.10	1.0 ± 0.14	4.02 ± 0.45	3.19 ± 0.24	8.61
g C m^{-2} day^{-2}	0.2	0.5	2.01	1.595	4.31
Ratio (%)	4.6	11.6	46.7	37.0	
Secondary forest					
g biomass m^{-2} day^{-1}	0.61 ± 0.16	0.36 ± 0.12	2.16 ± 0.32	4.13 ± 0.01	7.26
g C m^{-2} day^{-2}	0.31	0.18	1.08	2.067	3.63
Ratio (%)	8.4	5.0	29.8	56.9	

Total NPP in old-growth forest was 8.61 g m^{-2} day^{-1} in which, aboveground litterfall contributed 46.7%, reduced to AGB increment (37%), to fine root production (11.6), and to coarse root increment (4.6%) while total NPP in secondary forest was 7.26 g m^{-2} day^{-1}. Among these, the highest contribution belonged to AGB increment (56.9%), reduced to aboveground litterfall (29.8%), to coarse root increment (8.4%), and to fine root production (5.0%).

Soil Respiration

Results show that soil respiration was higher in old-growth forest compared to that in secondary forest regardless of seasonality (see Fig. 4.7a). In term of seasonality, it was highest in July (summer) and lowest in December (winter) for both forest stands. Annual mean soil respiration was 3.6 g C m^{-2} d^{-1} in old-growth forest (see Fig. 4.7b), double that in secondary forest (1.8 g C m^{-2} d^{-1}).

Net Ecosystem Production

As shown earlier, there was a higher NPP (Table 4.1) in old-growth forest (15.71 Mg C ha^{-1} y^{-1}), compared to that in secondary forest (13.25 Mg C ha^{-1} y^{-1}). However, soil respiration was much higher in old-growth forest (13.14 Mg C ha^{-1} y^{-1}), compared to that in secondary forest (6.68 Mg C ha^{-1} y^{-1}). These led to much higher NEP in secondary forest (6.57 Mg C ha^{-1} y^{-1}), compared to that in old-growth forest (2.57 Mg C ha^{-1} y^{-1}).

The accuracy of NEP estimations depends on methods used to estimate each component of NPP and to measure soil respiration. The biggest challenge is to estimate fine root production more accurately. Most previous studies have ignored decomposition in estimating fine root production (Tadaki 1968; Yashiro et al. 2010; Baishya and Barik 2011; Tateno et al. 2004; Girardin et al. 2010), leading to

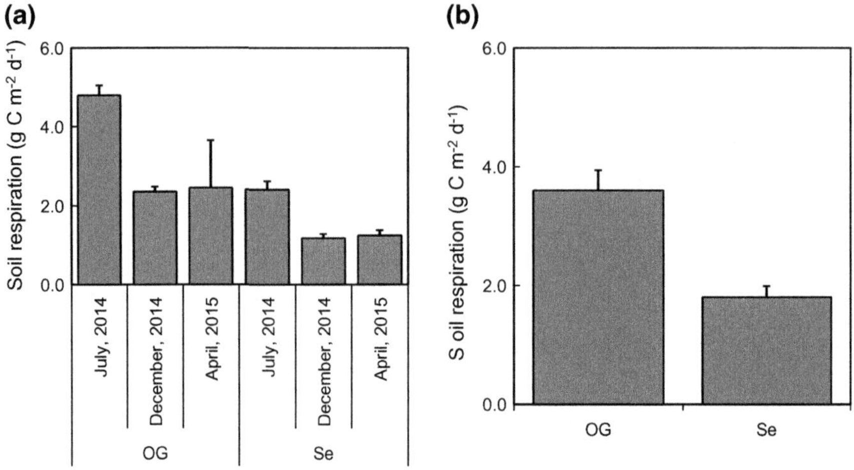

Fig. 4.7 Seasonal variation (**a**) and annual mean (**b**) of soil respiration. OG is old-growth forest, Se is secondary forest

underestimates. Meanwhile, it is important to note that applying a continuous inflow method in the present study is time-consuming and has high labor costs. In addition, the accuracy of soil respiration measurements depends much on equipment. Simple equipment was utilized in the present study and thus, there is a strong probability it may lead to higher error when compared to the utilization of other equipment (Ohtsuka et al. 2007, 2013).

As noted below, NEP is forest-type and age-dependent (see Table 4.3). Generally, it is higher in mixed forests than in pure forests and is higher in young forests than old ones. NEP of old-growth forest in the present study is similar to that in other forests as in the cool-temperate deciduous broad-leaved forests found in Japan (Saigusa et al. 2005; Ohtsuka et al. 2007). Meanwhile, the secondary forest in the present study has NEP similar to that in conifer forests in the UK (Medlyn et al. 2005). It is obvious that old-growth forest in the present study probably is not a primary forest, since NEP is still high as 2.6 Mg C ha^{-1} y^{-1}. In addition, previous research (Table 4.3) has shown that primary forests in the Pasoh Forest Reserve, Malaysia (−1.9 Mg C ha^{-1} y^{-1}) and a boreal black spruce (*Picea mariana*) forest in Canada (−1.3 Mg C ha^{-1} y^{-1}) have negative NEP. These indicated that NEP estimation should be conducted locally in order to produce a concrete basis for carbon price identification.

Conclusion: Carbon Market and Its Price for Vietnam

The international carbon market was developed under the Kyoto Protocol, and today carbon pricing initiatives are proposed, continuously, at regional and national levels, especially in developing countries. This alone underlines the strong endorsement that carbon pricing still receives positive attention, compared to other policy instruments to reduce greenhouse gas emissions (World Bank 2013). Like

Table 4.3 Net ecosystem production (NEP) of different forests around the world

Forest type	Study location	NEP ($Mg\ C\ ha^{-1}\ y^{-1}$)	Measuring period	Sources
Evergreen old-growth broad-leaved forest	21°23'N, 103°38'E (Northwest Vietnam)	2.6	2014–2015	The present study
Evergreen secondary broad-leaved forest		6.6		
A boreal black spruce (*Picea mariana*) forest	54°N, 105°W (Saskatchewan, Canada)	−1.3		O'Connell et al. 2003
A cool-temperate deciduous broad-leaved forest	36°08'N, 137°25'E (Central Japan)	2.1	1999–2003	Ohtsuka et al. 2007
		2.4	1994–2003	Saigusa et al. 2005
Conifer forests	64°07'N, 19°27'E (Flakaliden, Sweden)	0.7	1997–1998	Medlyn et al. 2005
	44°42'N, 0°43'W (Bray, French)	2.7	1997–1998	
	56°37'N, 3°48'W (Griffin, U.K.)	6.1	1998	
A young larch forest	62°13'N, 129°10'E' (Yakutsk, Russia.)	2.4	1998–1999	Sawamoto et al. 2003
A Japanese red pine (*Pinus densiflora*) stand	35°27'N, 138°46'E, (Mountain Fuji, Japan)	2.9	1999–2008	Otsuka et al. 2013
A 40-year-old Japanese cedar (*Cryptomeria japonica*) plantation	36°08'N, 137°22'E	4.3	2005–2009	Yashiro et al. 2010
Jack pine and aspen stands	55°51'N, 98°29'E (Manitoba, Canada)	1.0	2003	Goulden et al. 2011
Dense stand of 500 cm tall jack pine		1.7	2003	
A primary rainforest	2°5'N, 102°18'E (The Pasoh Forest Reserve, Malaysia)	−1.9	2004	Adachi et al. 2011
A dry evergreen forest	14°30'N, 101°55'E (The Sakaerat Environmental Research Station, Thailand)	0.7	2004	

Negative values (–1.3 and –1.9) in column NEP indicate no carbon accumulation

any other product, the carbon price fluctuates year by year, region by region, and country by country, and as noted by many scholars, it is higher in Europe and developed countries such as Japan or Australia than many other parts of the world (Peters-Stanley and Yin 2013; Davide 2014).

Several national and regional emission trading schemes are already in place, such as the EU ETS, the 2008 New Zealand ETS, and the Regional Greenhouse Gas Initiative. However, such schemes have not yet been established in many parts of the world (Davide 2014). The carbon price is implemented in the form of carbon taxes to CO_2 emitters. On the other hand, CO_2 emitters can buy carbon certificates from CO_2 absorbers such as forest owners and/or protectors to ensure that they pay for the same amount of CO_2 they emitted to the atmosphere. It is obvious that to gain a carbon certificates, forest owners and/or protectors must approve how much carbon is now stored and will be accumulated in their own forests. This is not an easy work especially in developing countries, where human resources, research equipment, and scientific basics, estimating technology, etc. are still limited. Much remains unclear over those who are deemed to be responsible for issuing carbon certificates.

For many developing countries including Vietnam, a government body is now responsible for issuing carbon certificates to forest owners. Unfortunately, the process is quite complicated and time-consuming. Moreover, most CO_2 emitters are government bodies. The government is not willing to issue carbon certificates and to pay for their own CO_2 emissions and as such many issues are taken up and dealt with by society-based NGOs whose work are nonprofit but have the rights to issue carbon certificates to forest owners/protectors.

Like many neighboring countries in the Southeast Asian region, forest management in Vietnam is complicated and confusing for local people. They are legally acknowledged to have protected natural forests and receive "protection fees" from the government and with that, they are not allowed to log timber. The main issue is, however, how much should the government pay for local people/forest protectors to ensure the sustainable management of protected forest? Likewise, local people want to receive as much as they are entitled to, at least to minimally sustain daily life, not only to survive above the poverty level. Notwithstanding the government tries to pay as little as possible. Therefore, identifying feasible and effective pricing of forest carbon is necessary for sustainable natural forest management.

It is in this socioeconomic context that this research aims to make contributions. Two types of carbon in natural forests, which have a long history of growth, must be considered in identifying payments to protectors. First, carbon stock is the amount of carbon currently stored in a forest ecosystem. Second, the amount of carbon will be accumulated (NEP) in forest ecosystem. Without protection efforts, forests will be disturbed by illegal logging, which eventually leads to a reduction of carbon stock and NEP. Therefore, payments must cover both efforts to protect carbon stock and that to increase carbon accumulation (NEP). This means that we require adequate working knowledge of both carbon stock and NEP in any natural forests, which are protected by local people.

Identifying carbon prices is not easy work (MRP 2014). It must be based on the economic status and conditions of each country, as there exist higher prices in developed countries and lower prices in developing ones (Masood 2013). For sustainable natural forest management, the rule of thumb is that forest carbon pricing must ensure the minimal daily life of protectors. Furthermore, protecting forests not only protects their carbon stock and capacity to accumulate carbon, but also conserve biodiversity, and reduce natural disasters (such as landslides, flooding, etc.). All of these are tangible to the whole ecosystem and such services must be included in forest carbon pricing. An example is that to successfully protect

Southeast Asian birds in Sabah, Borneo island, and Malaysia, the carbon price is set as US $28 per ton of carbon for primary forest and it is even higher at US $47 per ton of carbon for a mixed forest of timber and palm (Edwards et al. 2012).

In Vietnam, carbon price should be identified locally based on economic and natural status other than using one flat rate for the whole country. This ensures that payment for carbon must minimally sustain the livelihood of local people. For example, in the northwest region of Vietnam, local people sustain their life through paddy rice and shifting cultivation, leading to low requirements for daily life. In the Central Highland region, growing coffee brings much higher income for local people. If the same rate is paid to local people, who live in Central Highland, they cannot sustain their life and this can lead to illegally logging forests: unsustainable forest management. Until recently, there has not been any research on carbon prices in Vietnam to know how much carbon price should be applied for minimally sustaining the daily life of local people. For example, a six-person household in the northwest region participates in protecting 10 ha of secondary natural forests, equaling 65 tons carbon accumulated in their forests per year (Table 4.1). To minimally sustain their life, it requires 1.080 kg of rice (15 kg rice per person per month), equaling US $648 (0.6 US $/1 kg rice). Then, the carbon price must be 10 US $/ton carbon. Current payment for forest protection in Vietnam is US $10 ha/year without considering carbon accumulation. Therefore, US $55 ha/year more should be paid to forest protectors and this should come from CO_2 emitters through forest carbon certificates. Local people can get a forest carbon certificate from responsible authorities and sell it to CO_2 emitters. A price of US $10/ton carbon in Vietnam is reasonable, compared to a higher price in Malaysia (Edwards et al. 2012).

Acknowledgments This research was funded by Vietnamese National Foundation for Science and Technology Development (NAFOSTED), under grant number 106-NN.06-2016.10.

Appendix: Methodologies

Data Collection

One plot of 30 m × 30 m was established in old-growth forest and another plot of the same size was established in 35-year-old secondary forest for NEP estimation. In each plot, all stems with DBH (diameter at breast height) $\geq$ 5 cm were measured and identified to species level in March 2014 and April 2015 for ΔM estimation. In each 30 m × 30 m plot, fifteen plots of 1 m × 1 m each were established in March 2014 on a forest surface floor by removing all materials for litterfall collection. Litter was collected three times in July, December 2014, and April 2015. Sequence soil cores were collected in March 2014 and on the same dates of litter collection. On each date, 30 soil cores were collected using stainless steel tube of 36 mm in diameter (inner diameter of 34 mm), and coring to a depth of 21 cm. 15 litter bags were buried in March 2014 and collected in July 2014.

Fig. 4.8 Litter trap (a), soil core sampling (b), and measuring soil respiration (c)

Another 15 bags were buried in July 2014 and were collected in December. Finally, 15 litter bags were buried in December 2014 and were collected in April 2015. Then, γ_{ij} (decomposition ratio) was estimated for corresponding intervals (March −July 2014, July−December 2014, December 2014−April 2015). Fine root production was then estimated by continuous inflow method in Eq. 4.6.

Five close chambers (see Fig. 4.8c) were set up in each 30 m × 30 m plot for soil respiration (R_s) measurement. R_s was measured on the same dates as collecting litterfall and sequence soil core sampling. Measuring CO_2 concentration in ppm in chambers was conducted continuously through days including the night. Temperature and moisture inside chambers were also measured parallel with R_s. Through CO_2 concentration in chamber, it was converted to mass of carbon.

Estimation Methodology

Net ecosystem production (NEP) or rate of carbon accumulation in a forest ecosystem is simply estimated in Eq. 4.1 (O'Connell et al. 2003).

$$NEP = NPP - R_s, \tag{4.1}$$

where NPP is net primary production and R_s is heterotrophic respiration (soil respiration).

NPP Estimation

There are four compartments included in NPP following Eq. 4.2.

$$\mathrm{NPP} = \Delta \mathrm{M} + \Delta \mathrm{Cr} + \mathrm{Lf} + \mathrm{Pr}, \tag{4.2}$$

where ΔM is aboveground biomass increment, ΔCr is coarse root increment, Lf is aboveground litterfall, and Pr is fine root production.

Aboveground Biomass Increment

ΔM is estimated in the following manner. Diameter at breast height (DBH) of all living stems is measured at time t_i and t_j ($t_j > t_i$). Then, aboveground biomass increment is estimated in Eq. 4.3 (Clark et al. 2001).

$$\begin{aligned}\Delta \mathrm{M} = \sum_{\mathrm{stem1}}^{\mathrm{stemn}} \mathrm{AGB}_{\mathrm{stema_{intj}}} - \mathrm{AGB}_{\mathrm{stema_{inti}}} + \sum_{\mathrm{stem1}}^{\mathrm{stemm}} \mathrm{AGB}_{\ \mathrm{ingrowthstemb_{intj}}} \\ - \mathrm{AGB}_{\mathrm{ingrowthstemb_{atDBH=5cm}}}\end{aligned} \tag{4.3}$$

Aboveground biomass (AGB; kg dry weight) of each stem is estimated based on allometry. Generally, a site-specific allometry results in more accurate AGB estimation than non-site-specific one. However, destructive sampling a number of trees is required and it is costly for allometry establishment, worldwide allometry is usually accepted in application following Eq. 4.4 (Chave et al. 2005).

$$\mathrm{AGB} = \rho * \exp\left[\begin{matrix} -1.499 + 2.148\ln(DBH) + \\ 0.207(\ln(DBH))^2 - 0.0281(\ln(DBH))^3 \end{matrix}\right], \tag{4.4}$$

where ρ is wood-specific gravity, and DBH is diameter at breast height in cm.

Coarse Root Increment

ΔCr is estimated based on relationship between coarse root biomass (CRB) and AGB (Mokany et al. 2006) following Eq. 4.5.

$$\mathrm{CRB} = 0.489\mathrm{AGB}^{\wedge 0.890} \tag{4.5}$$

Equation 5 is a worldwide equation, therefore, it can be applied to any tropical forests. If we know AGB at time t_j and t_i through its DBH, respectively, then $\Delta \mathrm{Cr} = \mathrm{CRB}_j - \mathrm{CRB}_i$.

Aboveground Litterfall

Lf (including all falling materials as leaves, branches, productive organs, etc.) is estimated by litter trap technique (see Fig. 4.8a). Litter traps made by cloth of circle or square shape are widely used. They are set up generally 1 m from surface ground and litterfall is collected periodically. For much research in developing countries, as local people usually disturb litter traps by collecting cloth, plots of square shape (1 m × 1 m) established on forest surface floor by removing all living and dead materials are usually used for reducing cost and disturbance. The shape of litter traps, materials used, and collected intervals vary by case. There are no specific requirements.

Fine Root Production

There are several methods for estimating fine root production (e.g., Osawa and Aizawa 2012; Metcalfe et al. 2007; Majdi et al. 2005; Bernier and Robitaille 2004; Hendricks and Pregitzer 1993; Raich and Nadelhoffer 1989; Fairlay and Alexander 1985), all having advantages and disadvantages. Continuous inflow method (Osawa and Aizawa 2012) is introduced, since it is a new method, results in a highly accurate estimation, and uses rather simple techniques as sequence soil core sampling and litter bag techniques.

Pr (fine root production) is estimated in Eq. 4.6.

$$Pr = \left(B_j - B_i\right) + \left(N_j - N_i\right) + \left[-\left(N_j - N_i\right) - \left(\left(N_j - N_i\right)/\gamma_{ij} + N_i\right) * \ln\left(1 - \gamma_{ij}\right)\right], \quad (4.6)$$

where B_i and B_j are mass of living fine roots (biomass) at time t_i and t_j, respectively ($t_j > t_i$), N_i and N_j are mass of dead fine roots (necromass) at time t_i and t_j, respectively, and γ_{ij} is decomposition ratio of dead fine roots between t_i and t_j. Sequential soil core sampling was used for B_i, B_j, N_i, and N_j, while litter bag technique was used for γ_{ij}.

For sequential soil core sampling (see Fig. 4.8b), stainless steel tubes of 36 mm in diameter (inner diameter of 34 mm) were used to core the ground sequentially to the depth of 21 cm. Fine roots are separated by washing and sieving collected soil. Then, the fine roots were classified into dead and live roots by their color, resilience, and structural integrity (Hishi and Takeda 2005). Roots were then dried in a forced-air oven at 70°C until constant mass, and the masses of live roots and dead roots were weighed separately.

The litter bag technique is used for γ_{ij}. Envelope-type root litter bags made from special cloth (root-impermeable water-permeable sheet/RIWP, Toyobo Co., Osaka, Japan) were used. The RIWP has a pore size of approximate 0.6 μm, which blocks practically the ingrowth of fine roots; however, fine soil particles, water, and microorganisms can penetrate through the sheet. Dead fine roots are collected from the field, washed free of soil, and then oven dried at 70°C until constant mass. Approximate one gram of fine roots is put into a litter bag (10 cm × 10 cm). Litter bags are buried to 10–15 cm soil depth at time t_i, then, are collected at time t_j.

Collected litter bags are washed, and then oven dried for remained mass. γ_{ij} is estimated as equaling (initial mass—remained mass)/initial mass.

Converting NPP to Carbon

To convert NPP to carbon, the dry mass containing 50% carbon was assumed (Hoen and Solberg 1994; Sarmiento et al. 2005).

Soil Respiration Measurement

Similar to other field research, we conducted soil respiration measurement. As one component to estimate NEP, soil respiration is important to be conducted well on site. Soil respiration (R_s) by microorganisms to decompose litter in soil and on forest floor is known as heterotrophic respiration. This process returns nutrients to soil for sustaining the life of plants, however, it also emits a huge amount of CO_2 to the atmosphere (Schlesinger and Andrews 2000). A closed chamber method (CC-method) using an infrared gas analyzer—IRGA (Bekku et al. 1995) has been widely used in estimating soil respiration.

Soil CO_2 efflux includes autotrophic respiration from living roots (R_a) and heterotrophic respiration (R_s) from microbes and soil fauna to decompose organic matter. For estimating NEP, R_s must be known. R_s is measured using close-chamber method (Bekku et al. 1995). To separate R_a from soil CO_2 efflux for measuring R_s, plot is made (see Fig. 4.8c) by trenching to a depth of 80 cm. This ensures no roots survive inside the plot, therefore, soil CO_2 efflux from that plot contains only R_s.

Statistical Analysis

Spatial variation in ΔM and ΔCr was quantified as standard errors. First, a 30 m × 30 m plot was divided into four 15 m × 15 m plots. Aboveground biomass in March 2014 and April 2015 was estimated for four 15 m × 15 m plots separately. The difference of biomass between 2014 and 2015 (a one-year basis) was then calculated for four 15 m × 15 m plots separately. Mean of biomass difference and its standard error from four 15 m × 15 m plots represented ΔM and its standard error. The same process was applied for ΔCr and its standard error. Spatial variation in Lf was quantified as a standard error and was estimated in the following manner: in each collection, the mean of Lf from 15 traps and its standard error represented values of the collected intervals. While mean and standard error from 15 traps of aboveground litterfall in all three collections were annual mean and its standard error.

For fine root production estimation, 30 cores in each collection date were randomly divided into six groups of five cores each. Means of five cores were used to estimate production, mortality, and decomposition by continuous inflow method in Eq. 4.6. The outputs were six values. Mean and its standard error of these six values were means of production, mortality, and decomposition with their standard errors.

Soil respiration was estimated as daily mean and its standard error from data of all recorded dates in July, December 2014, and April 2015.

NPP was first estimated as dry biomass (Mg ha^{-1} y^{-1}), then was converted to carbon (Mg C ha^{-1} y^{-1}) as equaled to 50% of dry biomass (Hoen and Solberg 1994, Sarmiento et al. 2005). Meanwhile, R_s was measured as ppm (part per million), then was also converted to carbon (Mg C ha^{-1} y^{-1}). Finally, the difference of NPP and R_s is carbon accumulation (Mg C ha^{-1} y^{-1}).

References

Adachi, M., Ito, A., Ishida, A., Kadir, W. R., Ladpala, P., & Yamagata, Y. (2011). Carbon budget of tropical forests in Southeast Asia and the effects of deforestation: An approach using a process-based model and field measurements. *Biogeosciences, 8,* 2635–2647.

Baishya, R., & Barik, S. K. (2011). Estimation of tree biomass, carbon pool and net primary production of an old-growth *Pinus kesiya* Royle ex. Gordon forest in north-eastern India. *Annals of Forest Science, 68,* 727–736.

Bekku, Y., Koizumi, H., Nakadai, T., & Iwaki, H. (1995). Measurement of soil respiration using closed chamber method: An IRGA technique. *Ecological Research, 10,* 369–373.

Bernier, P. Y., & Robitaille, G. (2004). A plane intersect method for estimating fine root productivity of trees from minirhizotron images. *Plant and Soil, 265,* 165–173.

Caspersen, J. P., Pacala, S. W., Jenkins, C. J., Hurtt, G. C., Moorcroft, P. R., & Birdsey, R. A. (2000). Contributions of land-use history to carbon accumulation in U.S. forests. *Science, 290,* 1148 –1151.

Chave, J., Andalo, C., Brown, S., Cairns, M. A., et al. (2005). Tree allometry and improved estimation of carbon stocks and balance in tropical forests. *Oecologia, 145,* 87–99.

Clark, D. A., Brown, S., Kicklighter, D., Chambers, J. Q., Thomlinson, J. R., & Ni, J. (2001). Measuring net primary production in forests: concepts and field methods. *Ecological Application, 11,* 356–370.

Davide, M. (2014). Carbon price to survive the crisis. *International Climate Policy & Carbon Markets, 30,* 2–5.

Edwards, D. P., Fisher, B., Wilcove, D. S., Edwards, F. A., Davies, R., & Hamer, K. (2012). Can payments for ecosystem services protect Southeast Asian birds? *BOU Proceedings – Ecosystem services: do we need birds?*

Fairley, R.I., & Alexander, I. J. (1985). Methods of calculating fine root production in forests. In A. Fitter, H. Atkinson & D. J. Read (Eds.) *Ecological interactions in soil: Plants, microbes and animals* (pp. 37–42). Oxford: Blackwell Scientific Publications.

Fukushima, M., Kanzaki, M., Hara, M., Ohkubo, T., Preechapanya, P., & Choocharoen, C. (2008). Secondary forest succession after swidden cultivation in the montane forest area in Northern Thailand. *Forest Ecology and Management, 255,* 1994–2006.

Girardin, C. A. J., Malhi, Y., Aragao, L. E. O. C., Mamani, M., Huaraca, H. W., Durand, L., et al. (2010). Net primary productivity allocation and cycling of carbon along a tropical forest elevational transect in the Peruvian Andes. *Global Change Biology, 16,* 3176–3192.

Goulden, M. L., Mcmillan, A. M. S., Winston, G. C., Rocha, A. V., Manies, K. L., Harden, J. W., et al. (2011). Patterns of NPP, GPP, respiration, and NEP during boreal forest succession. *Global Change Biology, 17,* 855–871.

Hendricks, R. L., & Pregitzer, K. S. (1993). The dynamics of fine root length, biomass, and nitrogen content in two northern hardwood ecosystems. *Canadian Journal Forest Research, 23,* 2507–2520.

Hishi, T., & Takeda, H. (2005). Dynamics of heterorhizic root systems: protoxylem groups within the fine-root system of *Chamaecyparis obtusa*. *New Phytologist, 167,* 509–521.

Hoen, H., & Solberg, B. (1994). Potential and economic efficiency of carbon sequestration in forest biomass through silvicultural management. *Forest Science, 40,* 429–451.

Majdi, H. K., Pregitzer, K., Moren, A. S., Nylund, J. E., & Agren, G. I. (2005). Measuring fine root turnover in forest ecosystems. *Plant and Soil, 276,* 1–8.

Masood, E. (2013). *Mapping carbon pricing initiative*. Carbon Financial at the World Bank.

Medlyn, B. E. P., Bobinson, R., Clement, et al. 2005. Carbon balance of coniferous forests growing in contracting climates: Model-based analysis. *Agricultural and Forest Meteorology, 131,* 97–124.

Metcalfe, D. F., Meir, P., & Williams, M. (2007). A comparison of methods for converting rhizotron root length measurements into estimates of root mass production per unit ground area. *Plant and Soil, 301,* 279–288.

Mokany, K., Raison, R. J., & Prokushkin, A. S. (2006). Critical analysis of root: shoot ratios in terrestrial biomes. *Global Change Biology, 12,* 84–96.

MRP. (2014). *Market readiness proposal*. Ministry of Natural Resources and Environment – Socialist Republic of Vietnam.

Nguyen, N. B. (1996). *Forest soil of Vietnam*. Hanoi: Agriculture Publishing House.

O'Connell, K. E. B., Gower, S. T., & Norman, J. M. (2003). Net Ecosystem Production of Two Contrasting Boreal Black Spruce Forest Communities. *Ecosystems, 6,* 248–260.

Ohtsuka, T., Mo, W., Satomura, T., Inatomi, M., & Koizumi, H. (2007). Biometric based carbon flux measurements and net ecosystem production (NEP) in a temperate deciduous broad-leaved forest beneath a flux tower. *Ecosystems, 10,* 324–334.

Ohtsuka, T., Negishi, M., Sugita, K., Iimura, Y., & Hirota, M. (2013). Carbon cycling and sequestration in Japanese red pine (*Pinus densiflora*) forest on lava flow of Mt. *Fuji*. *Ecological Research, 28,* 855–867.

Osawa, A., & Aizawa, R. (2012). A new approach to estimate fine root production, mortality, and decomposition using litter bag experiments and soil core techniques. *Plant and Soil, 355,* 167–181.

Peters-Stanley, M., & Yin, D. 2013. *Maneuvering the mosaic state of the voluntary carbon markets 2013*. A Report by Forest Trends' Ecosystem Marketplace & Bloomberg New Energy Finance.

Raich, J. W., & Nadelhoffer, K. J. (1989). Belowground carbon allocation in forest ecosystems: global trends. *Ecology, 70,* 1346–1354.

Randerson, J. T., Chaplin, F. S., Harden, J. W., Neff, J. C., & Harmon, M. E. (2002). Net ecosystem production: A comprehensive measure of net carbon accumulation by ecosystems. *Ecological Applications, 12,* 937–947.

Saigusa, N., Yamamoto, S., Murayama, S., & Kondo, H. (2005). Inter-annual variability of carbon budget components in an Asia-flux forest site estimated by long-term flux measurements. *Agricultural and Forest Meteorology, 134,* 17–26.

Sarmiento, G., Pinillos, M., & Garay, I. (2005). Biomass variability in tropical American lowland rainforests. *Ecotropicos, 18,* 1–20.

Sato, T., Kominami, Y., Saito, S., Niiyama, K., Tanouchi, H., Nagamatsu, D., et al. (2010). Temporal dynamics and resilience of fine aboveground litterfall in relation to typhoon disturbances over 14 years in an old-growth lucidophyllous forest in southwestern Japan. *Plant Ecology, 208,* 187–198.

Sawamoto, T., Hatano, R., Shibuya, M., Takahashi, K., Isaev, A. P., Desyatkin, R. M., et al. (2003). Changes in net ecosystem production associated with forest fire in Ihiga ecosystems, near Yakutsk, Russia. *Soil Science and Plant Nutrition, 49,* 93–501.

Schlesinger, H. W., & Andrews, A. J. (2000). Soil respiration and the global carbon cycle. *Biogeochemistry, 48,* 7–20.

Sherman, E. R., Fahey, J., & Martinez, T. (2003). Spatial patterns of biomass and aboveground net primary productivity in a mangrove ecosystem in the Dominican Republic. *Ecosystems, 6,* 384–398.

Tadaki, Y. (1968). Studies on the production structure of forest (XIV). The third report on the primary production of a young stand of *Castonopsis cuspidate*. *Journal of Japanese Forest Society, 50,* 60–64.

Tateno, R., Hishi, T., & Takeda, H. (2004). Above and belowground biomass and net primary production in a cool-temperate deciduous forest in relation to topographical changes in soil nitrogen. *Forest Ecology and Management, 193,* 297–306.

Tran, V. D., Don, L. K., & Hoang, V. T. (2005). Rehabilitation of the native forest tree species at the forest plantations and denuded hills of Namlau commune in Sonla Province. *Vietnam. Forest Science and Technology, 1,* 51–58.

Tran, V. D., Osawa, A., & Thang, N. T. (2010). Recovery process of a mountain forest after shifting cultivation in Northwestern Vietnam. *Forest Ecology and Management, 259,* 1650–1659.

Tran, V. D., Osawa, A., Thang, N. T., et al. (2011). Population changes of early successional forest species after shifting cultivation in Northwestern Vietnam. *New Forests, 41,* 247–262.

Whittaker, R. H., & Woodwell, G. M. (1968). Dimension and production relations of trees and shrubs in the Brookhaven forests, New York. *Journal of Ecology, 56,* 1–25.

World Bank. 2013. *Mapping carbon pricing initiatives*. Developments and prospects.

Wunder, S., Engel, S., & Pagiola, S. (2008). Payments for environmental services in developing and developed countries. *Ecological Economics, 65,* 834–852.

Yashiro, Y., Lee, N. Y. M., Ohtsuka, T., Shizu, Y., Saitoh, T. M., & Koizumi, H. (2010). Biometric-based estimation of net ecosystem production in a mature Japanese cedar (*Cryptomeria japonica*) plantation beneath a flux tower. *Journal of Plant Research, 123,* 463–472.

Chapter 5
Soil Information as a Reforestation Decision-Making Tool and Its Implication for Forest Management in the Philippines

I. A. Navarrete, D. P. Peque and M. D. Macabuhay

Abstract Over the last 2 decades, the Philippine government has devoted a considerable amount of resources to the rehabilitation and reforestation of degraded forest land. However, deforestation and forest degradation have continued to be a major environmental problem in the Philippines as vast forest cover is lost annually. While large-scale reforestation projects have been initiated, most have been far from successful. This chapter argues that soil is the single most important factor affecting survival, growth, and development of trees and thus, influence the success of reforestation programs. Within the context of the Philippines, it analyzes soil factors that limits the success of reforestation projects and discusses the use of exotic and native tree species in reforestation. Finally, it presents current research and development efforts to reforest degraded forest land, particularly the National Greening Program, and the role of the reforestation program on soil carbon sequestration.

Keywords Carbon sequestration · Degraded soil · Exotic species
Reforestation · Site quality

Introduction

Soil is one of the most important factors affecting the survival, growth, and development of trees. It also serves as a habitat for a multitude of soil organisms necessary for the nutrient cycling of elements in a forest ecosystem. Because of these, the sensible use of forest soil resources, particularly for reforestation pur-

I. A. Navarrete (✉) · M. D. Macabuhay
Department of Environmental Science, Ateneo de Manila University, Quezon City, Philippines
e-mail: inavarrete@ateneo.edu

D. P. Peque
Department of Forest Science, Visayas State University, Baybay, Philippines

M. Lopez and J. Suryomenggolo (eds.), *Environmental Resources Use and Challenges in Contemporary Southeast Asia*, Asia in Transition 7,
https://doi.org/10.1007/978-981-10-8881-0_5

poses, should be based on extensive knowledge on soil properties (e.g., Kauffman et al. 1998). For instance, there are lands that may be best left under forest to maintain an ecological balance because of the physical and chemical limitations of the soil make it unsuitable for agriculture. Several soil properties, including the depth of the litter layer, rooting depth, the mineral composition of the deeper subsoil, as well as the geomorphologic history of the landscape and spatial variability of soils at the site need to be understood when considering the conservation of natural forest vegetation and the biodiversity of these areas (Jahn and Asio 1998; Schlichting et al. 1995; Schneider et al. 2014). However, the role of soil as a component of forest ecosystem management has not yet been taken into consideration in forest management initiatives (e.g., reforestation) in the Philippines. For example, the common view of forest research management is focused on the growth performance, silvicultural practices, and pest interaction of trees, whereas soil is not considered because of the general perception that any tree can grow anywhere regardless of soil fertility. This chapter argues that the failure to recognize the importance of soil resources, particularly the properties of the soil in natural forest and plantation management had been, and still is, one of the major impediments in the development of the forestry sector in the country. The failure of most reforestation efforts in the past clearly indicates that forest restoration in degraded areas requires a novel or a holistic approach (Margraf and Milan 1996; Asio and Milan 2002).

Large areas devoted to large-scale reforestation programs across the country often lie in nutrient-poor and degraded lands developed mostly from sedimentary and volcanic parent materials (Asio et al. 2009; Navarrete et al. 2009). Several efforts in the past to rehabilitate these degraded lands used exotic species, particularly *Gmelina arborea, Tectona grandis, Swietenia macrophylla,* and *Eucalyptus* species (Margraf and Milan 1996; Asio and Milan 2002; Schneider et al. 2014). While there are a number of successful reforestation programs implemented, for example, the “Rainforestation Farming” initiative (Margraf and Milan 1996), many reforestation efforts done by government agencies in the past and even until today have failed (Pulhin et al. 2006). Several factors that led to these failures include poor tree seedling stock (Elliot et al. 2013), poor management and corrupt practices, poor site-tree specific matching (e.g., Langenberger et al. 2006), and the failure to consider soil properties (Schneider et al. 2014).

The aim of this chapter is to show how soil information is a fundamental component in any analysis of reforestation in the Philippines and should be a tool in reforestation programs. This chapter contains: (1) a brief history of deforestation in the Philippines, (2) past and current reforestation efforts, (3) the selection of tree species for reforestation, (4) the importance of increasing the knowledge of forest soils, (5) characteristics of forest soils, and (6) site qualities and soil constraints of forest soil. We further intend to provide an outlook of the role of soils in the success of the reforestation programs in the Philippines.

Deforestation in the Philippines

Before the Spaniards came to the Philippines in 1521, the country had an estimated 27.5 million hectares (ha) out of the 30 million ha (Fig. 5.1) of total primary forest (Wernstedt and Spencer 1967; Garrity et al. 1993). At the end of the Spanish colonial regime in 1899, forest cover was reduced to 21 million ha due to the great demand for timber for the construction of ships for the galleon trade (Bankoff 2007). The first modern logging operation was introduced by the Americans in 1904, when the Insular Lumber Company was granted a 20-year renewable concession to log the rich dipterocarp forest in Northern Negros, Visayas (Roth 1983). American colonial records show that the country lost 92,000 ha of forest cover annually from the period between 1900–1941 (Rebugio et al. 2007). The forestry sector further sustained and supported the country's economy after World War II, where the total area of primary forest declined rapidly to about 7.3 million ha (Garrity et al. 1993). By the 1980s, the annual rate of deforestation was estimated at 379,000 ha and was considered to be among the highest in developing countries (Kummer 1995; Garrity et al. 1993). By 1996, there was only 6.1 million ha of primary forest remaining, suggesting that in this last century alone, the country has lost about 15 million ha of primary forest (Lasco and Pulhin 1999).

Several studies considered shifting cultivation as the major cause of primary forest destruction after World War II due to commercial logging activities and the expansion of small-scale agriculture (Villavicencio 1987; Kummer 1995; Liu et al. 1993). Asio (1997) argued that population increase has changed the form and extent of upland agriculture, which lead to the degradation of the biophysical environment. For example, at the end of the last century, migration from the lowland to the upland areas, which started at a mere thousand, continuously increased to about 11 million by 1970 (Cruz 1984). This increased by 18 million in 1986 and an estimated 20 million or more at present. Cramb et al. (2000) observed that rural poverty has persuaded lowland farmers to migrate to the upland areas, where inappropriate cultivation practices resulted in soil degradation (Asio et al. 2009).

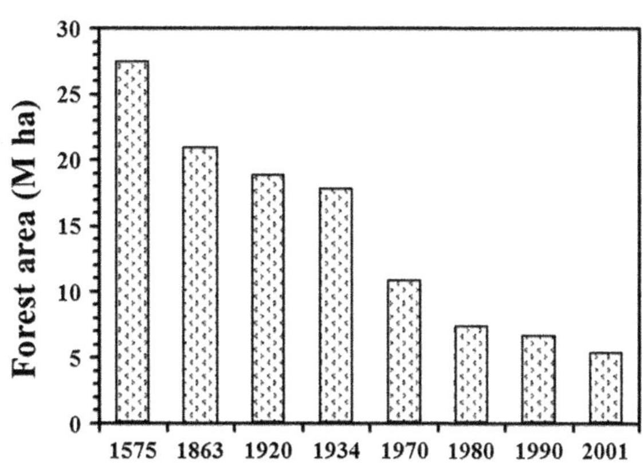

Fig. 5.1 Changes in the forest cover in the Philippines. (FMB 2003)

Liu et al. (1993) concluded that the presence of major roads is an important driver of deforestation and that the closer forests are on the roads, the higher the rate of deforestation. This is because roads created by logging concessionaires provide farmers with easy access to remaining primary forest. While logging causes more destruction than shifting cultivation (Liu et al. 1993), farmers usually follow unsuitable cultivation. Despite the differences in the views regarding the causes of deforestation, it is with certainity that primary forest cover in the country is decreasing at an alarming rate and is still under continuous threat even today.

As with many other islands in the Philippines, deforestation in the island of Leyte (7955 km^2), the eighth largest island in the archipelago, is also widespread (Kummer 1995). About, 45% (or 3578 km^2) of primary forest has been lost in the province in less than two decades from 1969 to 1984 (FMB 1988) due to illegal logging and land-use changes (Fig. 5.2). According to Kummer (1995), the primary forest cover of Leyte in the 1980s was <20% and recent estimates have shown that only 10% of the primary forest remain. Anticipating the worst ecological scenario, future forest management will focus on secondary forests. With forest loss, the unique fauna and flora diversity in the island, which is related to soil processes (Heemsbergen et al. 2004), is under increasing threat (Margraf and Milan 1996; Langenberger et al. 2006). This is apparent in the Philippines' ranking as second in the world's most threatened biodiversity hotspots (Myers et al. 2000). This is an indicator of the important role that biodiversity plays in the formation of soils in the country (Navarrete et al. 2007).

The major environmental effects of deforestation are: (i) soil erosion resulting in the removal of nutrient-rich topsoil and the siltation in rivers (Kummer 1995); (ii) loss of biodiversity due to the loss of habitat (Margraf and Milan 1996; Langenberger et al. 2006; Myers et al. 2000), and (iii) the release of greenhouse gases (e.g., CO_2) due to the burning of vegetation and unsustainable cultivation practices in the uplands, which contribute to global warming.

Lasco and Pulhin (1999) estimated that from the 1550s to the present, deforestation in the Philippines has released about 3.7 Gt C to the atmosphere, and of this amount, about 70% (2.6 Gt C) was released into the atmosphere in this century alone. This impact can be reduced if proper management of land is pursued, through the reforestation of degraded lands or the introduction of agroforestry systems that can increase carbon sequestration (e.g., Göltenboth and Milan 2015).

Past and Current Reforestation Efforts

Reforestation is not new in the Philippines. More than a century ago, the first Forestry School of the country (now College of Forestry and Natural Resources) established at Los Baños, Laguna in 1910, started a reforestation program (Pulhin et al. 2006). This was followed by various government-initiated projects aimed at bringing back the vegetative cover of denuded lands in the country caused by logging and shifting cultivation (Garrity et al. 1993). The earliest recorded

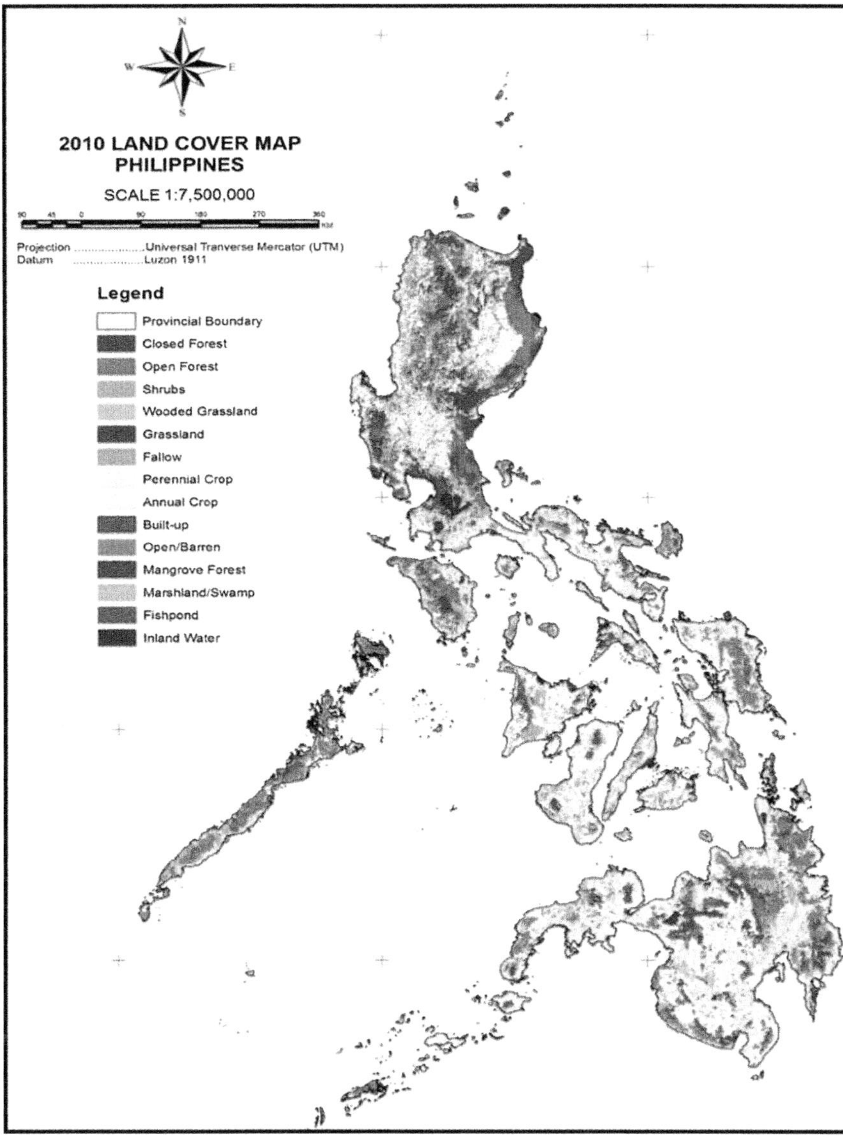

Fig. 5.2 Land cover map of the Philippines. (FMB 2012)

government reforestation program started during the American colonial period in 1916 in the Talisay Minglanilla Friar Lands Estate in Cebu Province. This was followed in 1919 by the Magsaysay Reforestation Project in Arayat, Ilocos, and Zambales areas, all on the island of Luzon (Rebugio et al. 2007). Earlier reforestation efforts were administered by the government until the mid-1970s, where various stakeholders, including farmers through social forestry, private and non-government organizations, and logging concessionaires were involved. From 1987–1992, the government embarked on a 5-year massive reforestation program, called the National Forestation Program (NFP), which targeted the reforestation of 300,000 ha of degraded lands. Despite the widespread reforestation projects initiated by the government, large areas of the country were still denuded (see, for example, Fig. 5.3), clearly suggesting that the past reforestation programs have failed. With current estimated forest cover of 6.84 million ha, the Philippines still has an estimated 8.21 million ha of degraded forest land out of the 15.05 million ha designated as forest lands (FMB 2012). From 1961 to 2012, the total area reforested in the country was already 2.2 million ha of which 350,323 ha or 15.3% was contributed by National Greening Program (NGP) (FMB 2012). Inspection on the ground, however, showed that the seedling survival rate is very low, which can be explained by a combination of technical and socioeconomic problems. The success of reforestation in the country was far from satisfactory (Israel and Lintag 2013) and overall success could be as low as 30% (Lasco and Pulhin 1999). This can be attributed to the limited protection of the reforested areas, limited after-planting maintenance and the lack of site-species matching (e.g., Schneider et al. 2014). Different stakeholders in the reforestation program should understand that not all species can grow well in any type of soil (Schneider et al. 2014). This is another justification of the importance of understanding the soil component to achieve success in any reforestation program.

Recently, the Philippines implemented the National Greening Program (NGP), a massive reforestation initiative, which aims to harmonize all tree planting efforts of the government sector, private organizations, and civil society. Launched on February 24, 2011 by virtue of an Executive Order No. 26 by President Benigno Simeon C. Aquino III, the NGP became a flagship program intended to address not only environmental degradation, food insecurity, biodiversity loss, and climate change but also poverty reduction in identified areas in the Philippines. NGP envisioned to plant 1.5 billion trees covering an area of about 1.5 million ha for a

Fig. 5.3 Typical degraded landscape in Maasin, Southern, Leyte, Philippines. *Photo* IA. Navarrete

period of six years, from 2011 to 2016. The NGP has four core components, namely: (i) survey, mapping, and planning (SMP), (ii) plantation establishment, (iii) seedling production or procurement, and (iv) protection of established plantations. Some of the activities under the first two components were founded on an ecological approach, namely: site characterization of the area to be reforested and the use of indigenous and native species of seedlings. The program recognized the need to adhere to some of the critical factors for plant survival, such as the soil type, elevation of planting sites, and climate. On the other hand, the NGP also encouraged the planting of native species due to their ability to adapt to local climate conditions, higher resistance to pests, and typhoons and their role in promoting diversity. However, these efforts can be deemed insufficient due to the incomplete assessment of the site, particularly the evaluation of the different characteristics of the soil, and the use of exotic species in reforestation activities. An analysis of the strength, weaknesses, opportunities and threat (SWOT) program (Fig. 5.4) shows that the inadequate integration of soil information on SMP and the absence of comprehensive monitoring and evaluation system in the NGP sites are among the factors that lead to the unsuccessful implementation of the program.

The aim of the NGP to plant 1.5 billion trees for a period of six years shows that the government-led reforestation initiative is target-oriented. The performance of the field offices is assessed based on the number of hectares planted with trees, and the suitability of the sites were not given much attention. Any available seedlings regardless of its genetic and phenotypic characteristics were planted to any identified forest lands, based on the belief that all trees can survive in any type of soil. This is further highlighted by a study conducted by the Philippine Institute for Development Studies, where they found that there is an incomplete assessment of the area to be reforested (Balangue 2015). Aside from the lack of emphasis on soil characterization, the NGP failed to determine site factors that can be considered as better predictors of survival rates of the seedlings planted (Balangue 2015). For

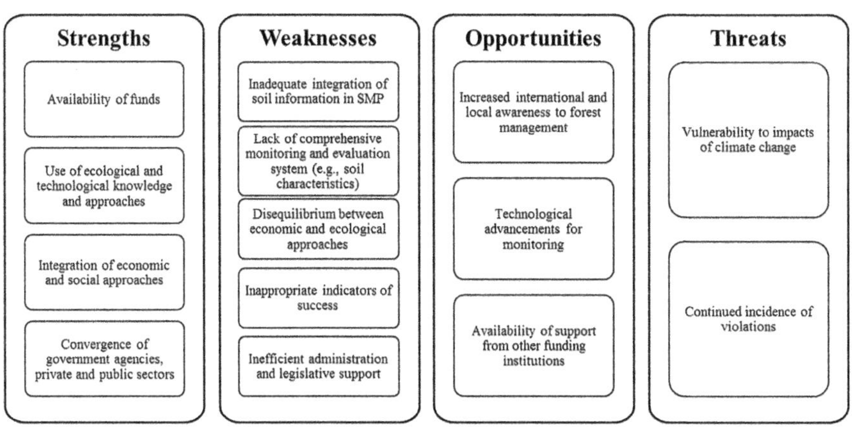

Fig. 5.4 Strength, weakness, opportunities and threats analysis of the National Greening Program

example, a study that will establish the soil moisture level per rainfall intensity in an area is important to establish the best indicator that will define the reforestation program in that particular place. Similarly, determination of appropriate temperature during the dry season is critical to the survival of forest plantations during the early stages of seedling development. As such, PIDS recommended that a full-blown in-depth forest restorability assessment of all degraded forestland situated in every barangay be conducted before embarking on a national reforestation program in the future. In 2013, the Philippine Commission on Audit reported that in 2012, NGP failed to achieve 85% survival rate of seedlings (Israel and Arbo 2015), which is no different from the lackluster performance of past reforestation efforts. While NGP reported that the target area to be planted has been achieved year-on-year, DENR has yet to publish the actual survival rates of the seedlings in NGP sites across the country, which cast uncertainty of the reported success of the NGP project.

Selection of Tree Species for Reforestation

Reforestation efforts in the Philippines are biased toward large-scale monoculture plantations of exotic species such as *G. arborea, T. grandis, A. mangium, S. macrophylla,* and *Eucalyptus* species (Margraf and Milan 1996; Dart et al. 2001; Asio and Milan 2002), which results to low success rates (Margraf and Milan 1996; Lasco and Pulhin 1999). Exotic species are preferred over native species in large-scale reforestation efforts because they are perceived to grow faster than native species (Table 5.1), their silviculture is better understood, their germplasm is widely available, and they have wide adaptability and tolerance to stress (Dart et al. 2001). However, the failure of exotic species to grow well in nutrient-poor and acidic soil can be attributed to the lack of site selection and site requirements of tree species (Dart et al. 2001; Schneider et al. 2014). This demonstrates that the growth and survival of native tree species are superior over exotic species because the former have the ability to grow in nutrient-poor soils with less silviculture management (Margraf and Milan 1996; Schneider et al. 2014). For example, Schneider et al. (2014) assessed the growth performance of 44 native and 16 introduced species in 25 reforestation sites in Leyte. They reported that some native species such as *Melia dubia* and *Terminalia microcarpa* had statistically higher mean annual height and diameter increments than *S. macrophylla*, whereas dipterocarps *Shorea guiso, S. contorta,* and *Parashorea malaanonan* all have statistically similar height with *S. macrophylla*. These results suggest that dipterocarps can also grow as fast as exotic species, and at the same time improve our understanding of native and exotic species for future reforestation projects (Schneider et al. 2014). While native species are encouraged in reforestation projects (Margraf and Milan 1996), the adoption of native species present difficulties in the reforestation program (Schneider et al. 2014). Hence, more studies are needed to determine the

Table 5.1 Physical and chemical characteristics of important soils in the humid tropics. (selected data from Kauffman et al. 1998)

Soil characteristics	Ferrasols (*n* = 22)		Acrisols (*n* = 22)		Luvisols (*n* = 9)		Cambisols (*n* = 30)	
	tops	subs	tops	subs	tops	subs	tops	subs
Sand (%)	35	29	52	36	40	30	25	26
Clay (%)	49	52	26	40	35	40	46	40
Bulk density (g/cm^3)	1.1	1.2	1.3	1.4	1.2	1.3	1.1	1.2
pH H_2O (1:2.5)	4.8	5.0	4.8	4.8	6.4	5.9	5.3	5.5
pH KCl (1:2.5)	4.1	4.5	4.1	4.0	5.5	4.6	4.6	4.5
Organic carbon (%)	2.3	0.4	2.0	0.4	2.2	0.3	2.3	0.4
$CEC_{pH\ 7}$ (cmolc/kg)	8.8	4.0	9.9	6.6	22.7	25	19.3	14.9
Base saturation (%)	19	19	26	12	87	67	49	52
Exch. Al ($cmol_c$/kg)	1.4	1.1	1.5	2.2	0.0	0.3	0.1	0.0

Tops = topsoil 0–20 cm; subs = subsoil 70–100 cm

regeneration behavior, germplasm availability, silviculture requirements, and growth rates of native species (Schneider et al. 2014).

The greater attention toward the use of exotic species in large-scale reforestation and the failure of these reforestation programs in most areas in the Philippines suggest that reforestation needs a different approach (i.e., integration of ecological concept), if not a holistic one to watershed management (Margraf and Milan 1996). During the 1990s, the Visayas State University (VSU) and the Gesellschaft für Technische Zusammenarbeit (GTZ) established the VSU-GTZ Applied Tropical Ecology Program or "*Rainforestation Farming*," a community-based reforestation program, which promoted the use of native tree species such as *Pterocarpus indicus*, *Dipterocarpus spp., P. plicata*, *Shorea spp,.* and many others (Margraf and Milan 1996; Balzer 1998) in the rehabilitation of the Leyte watershed. The use of native tree species was based on the hypothesis that a farming system that is closer in its physical structure and species composition to the local rainforest is more suitable (Margraf and Milan 1996). Since then, the rainforest farming program has been replicated in many areas in the Philippines and has been declared by the Philippines Department of Environment and Natural Resources (DENR) in 2004 as one of the official reforestation methods (Pulhin et al. 2006). According to Müller-Edzards (1996), the development of ecologically sustainable land use methods is of utmost importance in the reforestation process, and this can only be achieved if the species composition of native forests is known and the process of natural regeneration is understood. The *Rainforestation Farming* approach appears to be effective not only in restoring the forest vegetation but also in improving soil quality and in protecting watersheds on degraded sites (Margraf and Milan 1996; Asio and Milan 2002). Müller-Edzards (1996) further argued that native tree species should be used instead of exotic tree species for reforestation in order to prevent severe leaching of nutrients from the soil.

The Need for More Knowledge on Philippine (Forest) Soils

Highly weathered forest soils are widespread in the Philippines (Zikeli et al. 2000; Navarrete et al. 2007, 2009), and are mostly found where large-scale reforestation programs are located (Schneider et al. 2014). While it occupies about 70% of the land area of the country, very little is known concerning their nature and properties (e.g., Navarrete and Asio 2014). Blum (1998) pointed out that the lack of basic understanding about the ecological relationship within an ecosystem results in a trial and error method of land use, leading to soil degradation, a process that lowers the capacity of soils to produce goods or services. A better understanding of the soil can be achieved by studying the ecological site quality, thus providing a sound management recommendations aimed at minimizing the negative ecological impacts of soil degradation (Navarrete et al. 2013).

Large uncertainty remains in our understanding of degraded upland, including highly weathered soils because limited studies have been conducted on them (Asio et al. 2009; Navarrete and Asio 2014). Most soil studies in the Philippines (e.g., Navarrete and Asio 2014) abound with agronomic and plant nutrition studies related to the alleviation of nutrients in degraded upland soils (Asio et al. 2009). This justifies the need for more pedological studies on Philippine soils for suitable land-use and management of these soils require detailed information on their soil properties (Asio et al. 2009; Navarrete et al. 2013). Over the recent decades, there has been few pedological studies that have analyzed and reported on volcanic soils (Otsuka et al. 1988; Asio 1996; Jahn and Asio 1998; Navarrete et al. 2009), rain forest soils (Asio 1996; Zikeli et al. 2000; Navarrete et al. 2007), limestone soils (Asio et al. 2006; Navarrete et al. 2011) and highly weathered soil (Navarrete et al. 2007; Calubaquib et al. 2016).

Until now, the reconnaissance soil survey, which classified Philippine soils into soil series and soil types, is the major source of soil information (Asio 1996). For example, the reports of Barrera et al. (1954) and Simon et al. (1975) on the island Leyte and Samar, respectively, are still widely used by agricultural technicians and researchers in these areas. The aforementioned soil reports, although helpful, do not provide sufficient and updated information about soils of Leyte and Samar. It is perceived that most of the areas in Leyte and Samar covered in the above-mentioned survey is now degraded due to the unsustainable land use changes, which could be different from the time they were studied. In fact, to our knowledge, only one study exists in the literature about the soil chemical and mineralogical properties of highly weathered soils in the island of Samar (Navarrete et al. 2007). Hence, the poor understanding of the many problem soils faces such as degraded land in the Philippines merits continued research investigation (e.g., Navarrete and Asio 2014).

Characteristics of Degraded Forest Soils

Tropical rain forest soils, such as those found in the Philippines, have been perceived to be acidic and infertile (e.g., Markewitz et al. 2004). However, such perception is surprising because most the highly-diversified forest ecosystems in the world, such as the Amazon rainforest, grow well in nutrient-poor soils (Sanchez 1976). This perception has been challenged recently (Sanchez and Logan 1992), based on the premise that because of its diversity and complexity, tropical rainforest soils could range from young fertile soils to more highly weathered infertile soils (Sanchez 1976; Richter and Babbar 1991; Sanchez and Logan 1992).

About 64% of humid tropical soils are composed of Acrisol, Arenosol, Cambisol, Luvisol, Ferralsol, and Podzol. Of these major soil groups, 57% is occupied by Acrisol and Ferralsol, which display different soil limitations (Kauffman et al. 1998). Because of the highly weathered nature of these soils, they are acidic (pH < 5.5), have low organic matter (OM), have moderate to high acidity, are phosphorus (P) deficient and have high P-fixing capacity (Table 5.1). These could be problematic when these degraded soils are used for agriculture and reforestation purposes. High rainfall and good drainage favor intensive weathering of soil, and the net leaching results in the net loss of basic cations such as Ca, Mg, Na, and K making the soil acidic (Eswaran et al. 1992). Under this condition, exchangeable Al dominates the soil exchangeable complex (Kamprath 1980), and the presence of this toxic level of soluble Al^{3+} can impair the growth of the trees. Moreover, the presence of the high amounts of Fe and Al oxides give some soils a net positive charge at acidic condition, thus making soils unable to retain cation (Uehara 1978). Hence, the bulk of nutrients stored in these nutrient-poor soils in highly weathered soils is mainly related to the OM coming from forest vegetation (Sanchez 1976). While most of these soils have excellent rooting depth (> 1 m soil depth), the effective root zone is limited due to compaction, Al toxicity, available P deficiency and elemental imbalance (Eswaran et al. 1992).

The physicochemical properties of soils vary depending on the parent material (Schneider et al. 2014; Navarrete et al. 2013) and land use history (Asio et al. 1998; Navarrete et al. 2013; Navarrete and Tsutsuki 2008). On the one hand, degraded soil derived from limestone parent materials have an alkaline pH, shallow soil, the presence of toxic substances, and mostly deficient in micronutrients (Asio et al. 2006; Navarrete et al. 2011). On the other hand, soils derived from Andesite volcanic rocks have very high clay content, high water holding capacity, compacted, and low aggregate stability (Navarrete et al. 2007). In general, most degraded soils in the Philippines have low OM, moderate to high acidity, low to deficient exchangeable Ca, Mg, Na, and K, deficient P, toxic levels of Al, compacted and have poor soil structure (Asio et al. 2009). Thus, management of these soils should address these limitations (e.g., Navarrete et al. 2013). In terms of site characteristics, most degraded areas are covered with grasses such as *Imperata cylindrica, Saccharum spontaneum, Axonopus compressus, Paspalum conjugatum,* and *Cyrtococcum accrescens*, and shrubs, particularly *Melastoma offine* and

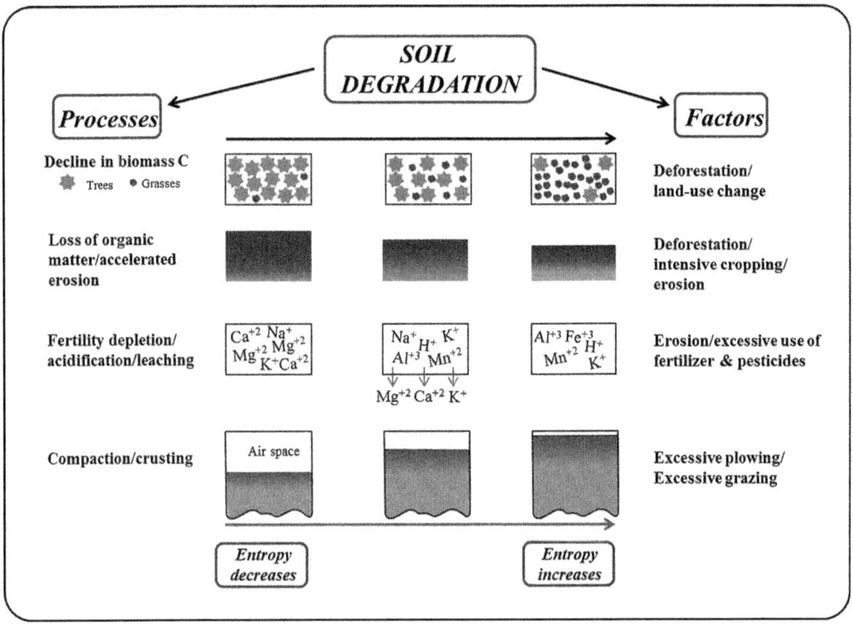

Fig. 5.5 Processes and factors of soil degradation in the Philippines

Psidium guajava (Asio 1996; Asio and Tulin 1997; Asio et al. 2009). Figure 5.5 illustrates the current soil degradation processes that are commonly occurring in most degraded upland soils in the Philippines.

Soil Qualities for Management

Drastic use and poor management of soil resources, specifically for reforestation and agriculture, can lead to soil degradation. Soil degradation affects the ecosystem services of soils, including plant production, buffering, transformation, filtering, cultural heritage, and infrastructure (Blum 1998). For example, the continuous use of degraded soils without good knowledge in terms of its physical and chemical properties has resulted in serious ecological problems (Alcala 1997).

Site qualities are collections of observable and measurable site characteristics (e.g., soil pH, OM, available P, and exchangeable basis), which have to be evaluated for specific land use (FAO 1976). The main purpose of determining the soil qualities is twofold. First, it seeks to present as clearly as possible the condition of an area. Second, it seeks to provide an accurate basis for the adoption of soils to trees or any crops. Accordingly, detailed information on soil properties and fertility status is needed to provide an accurate basis for the adoption of soils to crops needed for suitable land management. These include soil characteristics such as rooting depth, rootability, and water-, air-, energy and nutrient budgets (Schlichting et al. 1995). We argue here that reforestation efforts can only be successful if the ecological qualities of soils are properly evaluated, including the species composition of native forests and the process of natural regeneration. Such soil qualities require good data on soil characteristics (Table 5.2), which strongly influence the characteristics of the forest growing on it.

For example, the height of trees is affected by slope position, particularly for upper slope soil which has shallow solum. The shallow soil depth does not support the weight of tall and heavy trees, particularly for exotic species. Furthermore, shallow soils reduce rooting depth and correlate with the low water storage capacity that eventually results to water stress during dry periods (Baillie 1996). Whitmore (1998) noted that the availability of water is responsible for the occurrence and species composition of forest type and thus, serves as an important environmental factor limiting tree growth and distribution (Zimmermann and Brown 1971). Moreover, information about soil characteristics and ecological qualities are needed for understanding the total plant nutrient cycle in the natural ecosystem and the behavior of soil under various human influences. Finally, the information generated from site quality evaluation is important for biodiversity conservation. The ability of the soil to provide necessary plant nutrients and good environmental conditions determines the distribution of plant species in a specific area. Site requirements, such as habitat, parent material, and soil types, for a specific species is necessary for the establishment and distribution of trees to attain their full genetic potential (Table 5.3) (Langenberger 2000; Schneider et al. 2014). From these observations, many believed that on-site species trials are the only way to determine the species best-suited to a site.

Table 5.2 Site qualities of an Andosol and Alisol (Jahn and Asio 1998) and Ferralsol (Navarrete et al. 2007)

Soil qualities	Andosol (Ormoc)		Alisol (Baybay)		Ferralsol (Samar)	
	Value	Rating	Value	Rating	Value	Rating
Rooting depth	>100 cm	+++	>100 cm	+++	>100 cm	+++
Rootability		+++		++		+++
Erodability		±		–		–
Available water capacity	160 l/m^2	+++	140 l/m^2	+++	180 l/m^2	+++
Drainage		+++		++		+++
OM	42 kg/m^2		18 kg/m^2		13 kg/m^2	+
Available P	0.7 g/m^2	–	0.3 g/m^2	–	0.13 g/m^2	–
P retention	98%	+++	70%	++	69%	++
Available K	45 g/m^2	±	36 g/m^2	±	nd	nd
Available Ca	7 g/m^2	–	198 g/m^2	+	nd	nd
Available Mg	15 g/m^2	±	194 g/m^2	+++	nd	nd

nd: no data

Table 5.3 Soil requirements of some indigenous tree (Langenberger 2000) and selected Australian tree species (Harrison and Herborn 2000)

Tree Species	Soil characteristics
Indigenous Species	
Melia dubia (Bagalunga)	Growing in limestone soils, but could grow well in other soil types
Casuarina equisetifolia (Agoho)	Wide range of soils (acidic and alkaline)
Lagerstroema speciosa (Banaba)	Clay loam soil
Shorea contorta (White Lauan)	Deep, loamy to clayey volcanic soil
Vitex parviflora (Molave)	Dry rocky ridges; limestone soil; well-drained soils
Pterocarpus indicus (Narra)	Moist soil near gullies and stream banks; also in limestone soils
Dipterocarpus grandiflorus (Apitong)	Lowland well-drained soils; limestone soil
Parashorea malaanonan (Bagtikan)	Wide range of soils but prefers dry areas
Shorea guiso (Guijo)	Limestone soil; well-drained close to coast
Shorea negrosensis (Red Lauan)	Well-drained soils moist soil along river banks
Shorea falciferoides (Yakal-Yamban)	Well-drained soils on ridges and upper slopes
Australian Species	
Acacia auriculiformis	Wide range of soils (infertile, salt-affected, waterlogged)
Acacia mangium	Fertile soil
Eucalyptus camaldulensis	Wide range of soils even in infertile sandy soils
Eucalyptus grandis	Fertile, well-drained (alluvial or volcanic)
Eucalyptus tereticornis	Moist but not waterlogged soils with good nutrient availability (acid soils are not suited)

Soil fertility has a strong influence on tree species diversity (Korning et al. 1994; Kubota et al. 1998; Masunaga et al. 1998). The high diversity of trees in the western Amazonian rain forest is caused by local variations in soil characteristics (Korning et al. 1994). On the other hand, Kubota et al. (1998) reported that the high exchangeable Ca variations and the vertical and horizontal variations in the soil characteristics in the tropical forest soils of Indonesia are also imperative in supporting tree species diversity in the area. Masunaga et al. (1998) reported that the distribution characteristics of mineral elements in both trees and soil are the major factors controlling tree species communities. Using regression analysis of nine native species occurring with at least five individuals across five sites in Leyte, Schneider et al (2014) reported that several soil properties such as OM, nitrogen, pH, available P, clay content and cation exchange capacity showed a range of significant predictors for tree diameter growth in different tree species. The result suggests that some soil properties are good indicators of the growth of the trees and that many native species do better in certain soil types and site conditions, indicating a need for more refined planting guidelines (Schneider et al. 2014).

Improving the fertility of most nutrient-poor degraded soils, where most of the reforestation sites are located, is a big challenge. Fertility constraints in these soils could be overcome by the application of chemical fertilizers, particularly during the early stages of tree growth. However, frequently, the application of fertilizer does not show an improved growth of the plant (Zech and Drechsel 1998). This is not due to the lack of fertilizer usage, but due to the failure of the applied fertilizer to reach the root zone of young trees caused by the rapid leaching of the fertilizer beyond the root zone during heavy rain, adsorption onto strongly fixing soil surfaces and uptake by competing weedy species. Because of this, there is a need to develop more efficient and effective techniques for delivering nutrients to trees, which can be achieved by incorporation of long-term slow-release fertilizer into the potting medium with suitable properties that will give the plant a good start when transplanted into the field. However, one research gap is that little information is available to support, which nutrients are most limiting to the growth of dipterocarp seedlings planted in degraded soils (Nussbaum and Hoe 1996). Also, an approach to tree nutrition that will likely enhance growth of planted trees is the use of microorganisms, particularly ectomycorrhizal and vesicular arbuscular mycorrhiza, which enhanced symbiotic activity of the microorganism and the plant resulting in the increased availability of nutrients to the plants leading to good tree growth (e.g., Elliot et al. 2013).

Conclusion

The success or failure of any reforestation programs in the Philippines relies on the proper assessment of soil properties, particularly physical and chemical soil characteristics. In addition, information about site quality plays a significant role in the future management of the forest ecosystem. Without due field determination of soil properties, it is likely that any reforestation effort will be useless and that the resulting failure cannot be corrected promptly in the field. In addition, more research is needed for species-site matching, nursery management and silviculture management of native species plantations improve survival and performance (e.g., Schneider et al. 2014). Specifically, all tree species used in the reforestation program should be matched carefully to the site, not only to ensure the success of the reforestation effort but also to avoid any serious negative ecological impact of introduced species. There is a need to study the water and element dynamics in planted native species under different ecological conditions and management practices (Zech and Drechsel 1998).

However, many projects would benefit from increased coordination with other ongoing research activities. Building linkages and breaking disciplinary barriers among geology, soil science, agronomy, forestry, pasture management, hydrology, water resources management, agro-climatology, ecology, soil biology, and ecosystems research will improve the quality and applicability of research and provide new avenues for future integrated research and development programs.

Similar to reforestation activities, research into the social, political, and economic issues must also be tied to the biophysical aspects of research to provide a holistic approach to soil management. These research endeavors should inform policy-making bodies for the development of a roadmap for reforestation for the purpose of soil protection. The costs, benefits and expected impacts of soil management technologies and practices and intervention approaches, as well as the costs of no-action and the need for multidisciplinary approaches, must be taken into account in developing the roadmap for soil management and protection. Soils need greater investment in all fields, including raising awareness and mobilizing stakeholders and policy makers for the integration of soil information in reforestation.

The failure of the NGP, along with previous reforestation efforts carried out in the Philippines, highlights the need to increase the awareness of stakeholders, particularly the government, on the importance of soils vis-à-vis the reforestation efforts. Developed countries, such as European Union members, have acknowledged the finite nature of soils and the ramifications of the unsustainable use of soils to the quality and quantity of Europe's soil stock. The European Commission's Financial Instrument for the Environment (LIFE) follows two main guiding principles related to soil sustainability, namely (1) preventing further soil degradation and preserving its functions, and (2) restoring degraded soils to a level of functionality consistent at least with current and intended use (Camarsa et al. 2014). Similarly, Indonesia has taken a step forward in protecting its peatlands with the establishment of the Peatland Restoration Agency. Increased awareness on the role peat soils play in carbon storage came after a study conducted by the World Bank (2016) showed that clearing and draining of peatlands for oil palm and pulpwood plantations caused damage to the Indonesian economy, costing the nation twice as much as it took to rebuild after the 2004 tsunami. Hence, the evaluation of reforestation programs and the planning of new strategies should incorporate the potential economic implications of the failure to integrate soil protection in the reforestation efforts. Now is the time that government, particularly policy-making and implementing bodies, should recognize the issue of soil degradation in the country and that approaches to forest management, particularly the reforestation program, should be holistic and that soil be an integral part of the solution.

Acknowledgements The University Research Council of Ateneo de Manila University has granted a research faculty fellowship to Ian A. Navarrete during this preparation of the manuscript.

References

Alcala, A. C. (1997). Keynote address. In *Proceedings of international conference on reforestation with Philippine species*. ViSCA-GTZ Ecology Project, Baybay, Leyte, Philippines.

Asio, V. B. (1996). *Characteristics, weathering, formation and degradation of soils from volcanic rocks in Leyte, Philippines*. Hohenheimer Bodenkundliche Hefte Vol. 33 (p. 209). Stuttgart.

Asio, V. B. (1997). A review of upland agriculture, population pressure, and environmental degradation in the Philippines. *Annals of Tropical Research, 19,* 1–18.
Asio, V. B., & Milan, P. P. (2002). *Improvement of soil quality in degraded lands through rainforestation farming.* Paper presented during the International Symposium on Sustaining Food Security and Managing Natural Resources in Southeast Asia, January 8–11, 2002, Chiang Mai, Thailand.
Asio, V. B., & Tulin, A. B. (1997). *The characteristics of degraded soils and the ameliorative effect of vegetation.* Paper presented during the International Conference on Reforestation (p. 15). March 3–6, 1997, Tacloban City, Leyte.
Asio, V. B., Jahn, R., Stahr, K., & Margraf, J. (1998). Soils of the tropical forest of Leyte. 2. Impact of different land uses on the status of organic matter and nutrient availability. In A. Schulte & D. Ruhiyat (Eds.), *Soils of tropical forest ecosystems* (pp. 38–44). Springer-Verlag: Berlin.
Asio, V. B., Cabunos, C. C., & Chen, Z. C. (2006). Morphology, physicochemical characteristics, and fertility of soils from Quaternary limestone in Leyte, Philippines. *Soil Science, 171,* 648–661.
Asio, V. B., Jahn, R., Perez, F. O., Navarrete, I. A., & Abit, S. M., Jr. (2009). A review of soil degradation in the Philippines. *Annals of Tropical Research, 31,* 69–94.
Baillie, I. C. (1996). Soils of the humid tropics. In P. W. Richards (Ed.), *The tropical rainforest* (pp. 256–283). Cambridge: Cambridge University Press.
Balangue, T. O. (2015). National Greening Program assessment project: environmental component-process evaluation phase. *Philippine Institute for Development Studies.*
Balzer, P. (1998). *Dipterocarpus validus,* a native reforestation species. In F. Goltenboth, P. P. Milan & V. B. Asio (Eds.), *Proceeding of the International conference on applied tropical ecology: aspects on ecosystems management in tropical Asia.* September 8–10, 1998, VISCA, Baybay, Leyte, Philippines.
Bankoff, G. (2007). One island too many: Reappraising the extent of deforestation in the Philippines prior to 1946. *Journal of Historical Geography, 33,* 314–334.
Barrera, A., Aristorenas, I., & Tingzon, J. (1954). *Soil survey of Leyte province, Philippines* (p. 103). Manila: Department of Agriculture and Natural Resources (DANR).
Blum, W. E. H. (1998). Basic concepts: Degradation, resilience, and rehabilitation. In R. InLal & W. E. H. Blum (Eds.), *Methods for Assessment of Soil Degradation, Advances in Soil Science* (pp. 1–16). Boca Raton: CRC Press.
Bruijnzeel, L. A. (1998). Soil chemical changes after tropical forest disturbance and conversion: The hydrological perspective. In A. Schulte & D. Ruhiyat (Eds.), *Soils of Tropical Forest Ecosystems* (pp. 46–61). Berlin: Springer-Verlag.
Calubaquib, M. A. M., Navarrete, I. N., & Sanchez, P. B. (2016). Properties and nutrient status of degraded soils in Luzon, Philippines. *Philippine Journal of Science, 145,* 249–258
Camarsa, G., Silva, J., Toland, J., Hudson, T., Nottingham, S., Rosskopf, N., et al. (2014). *LIFE and soil protection* (p. 66). Luxembourg: Publications Office of the European Union.
Cramb, R. A., Garcia, J. N., Gerrits, R. V., & Saguiguit, G. C. (2000). Conservation farming projects in the Philippine uplands: Rhetoric and reality. *World Development, 28,* 911–927.
Cruz, M. C. (1984). Population pressure, migration and markets: Implications for upland developmend. In *Proceeding of the workshop on policies for forest resources management.* February 17–18, 1984, Club Solviento, Los Baños, Laguna.
Dart, P. J., Brown, S. M., Simpson, J. A., Harrison, S. R., & Venn T. J. (2001). Experience from ACIAR trials of the suitability and performance of Australian trees species. In S. Harrison & J. Herbohn (Eds.) *Socio-economic evaluation of the potential for Australian tree species in the Philippines* (p. 192). Canberra. ACIAR, Monograph 75. VIII.
Elliot, S., Blakesley, D., & Hardwick, K. (2013). *Restoring tropical forests: A practical guide* (p. 344). Richmond: Royal Botanic Gardan, Kew.
Eswaran, H., Kimble, J., Cook, T., & Beinroth, F.H. (1992). Soil diversity in the tropics: Implications for agricultural development. In R. Lal & P.A. Sanchez (Eds.), *Myths and science of soils of the tropics* (pp. 1–16). SSSA Special Publication No. 29.

FAO. (1976). A framework for land evaluation: Rome, Food and Agricultural Organization (FAO) of the United Nations, *Soils Bulletin, 32, Rome,* 72.
Forest Management Bureau (FMB). (2003). Sustainable forest management, poverty alleviation and food security in upland communities in the Philippines.
FMB. (2012). *Philippine forestry statistics.* Forest Management Bureau, Department of Environment and Natural Resources, Quezon City, Philippines.
FMB. (1988). *Natural forest resources of the Philippines.* Manila: Philippine-German Forest Resources Inventory Project.
Garrity, D. P., Kummer, D. M., & Guiang, E. S. (1993). The Philippines. In *Sustainable agriculture and the environment in the humid tropics* (pp. 549–624). Washington DC, National Academy Press.
Göltenboth, F., & Milan, P. P. (2015). Biological sinks for carbon dioxide-an option for agroforestry systems in the tropics. *Ann Trop Res, 37,* 129–141.
Harrison, S., Herbohn, J. (2000). *Socio-economic evaluation of the potential for Australian tree species in the Philippines* (p. 192). Canberra. ACIAR Monograph 75.
Heemsbergen, D. A., Berg, M. P., Loreau, M., Van Hal, J. R., Faber, J. H., & Verhoef, H. A. (2004). Biodiversity effects on soil processes explained by interspecific functional dissimilarity. *Science, 306,* 1019–1020.
Israel, D. C. & Lintag, J. H. (2013). *Assessment of the efficiency and effectiveness of the reforestation program of the Department of Environment and Natural Resources.* Philippine Institute for Development Studies.
Israel, D. C. & Arbo, M. D. G. (2015). *The National Greening Program: hope for our balding forests.* Philippine Institute for Development Studies.
Jahn, R., & Asio, V. B. (1998). Soils of the tropical forests of Leyte, Philippines. I. Weathering, soil characteristics, classification and site qualities. In A. Schulte & D. Ruhiyat (Eds.). *Soils of the tropical forest ecosystems* (pp. 29–36). Berlin: Springer-Verlag.
Kamprath, E. (1980). Soil acidity in well-drained soils of the tropics as a constraint to food production. *Priorities for Alleviating Soil-Related Constraints to Food Production in the Tropics* (pp. 171–187). Los Banos: IRRI.
Kauffman, S., Sombroek, W. G., & Mantel, S. (1998). Soils of rainforest: Characterization and major constraints of dominant forest soils in the humid tropics. In A. Schulte & D. Ruhiyat (Eds.), *Soils of the Tropical Forest Ecosystems* (pp. 9–20). Berlin: Springer- Verlag.
Korning, J., Thomsen, K., Dalisgaard, & Normber, P. (1994). Characteristics of three Udults and their relevance to the composition and structure of virgin forest of Amazonian Ecuador. *Geoderma, 63,* 145–164.
Kubota, D., Masunaga, T., Hermansah, A., Rasyidin, A., Hotta, M., Shinmura, Y., et al. (1998). Soil environment and tree species diversity in tropical rain forest, West Sumatra, Indonesia. In A. Schulte & D. Ruhiyat (Eds.), *Soils of the tropical forest ecosystems* (pp. 159–167). Berlin: Springer- Verlag.
Kummer, D. M. (1995). The political use of Philippine forestry statistics in the postwar period. *Crime, Law & Social Change, 22,* 163–180.
Langenberger, G. (2000). Forest vegetation studies on the foothills of Mt. Pangasugan, Leyte, Philippines. TOB Publication F-11/10e *Tropenokelogisches Begleitprogram,* GTZ, Eschborn, Germany.
Langenberger, G. (2006). Habitat distribution of dipterocarp species in the Leyte Cordillera: An indicator for species-site suitability in local reforestation programs. *Annals of Forest Science, 63,* 149–156.
Langenberger, G., Martin, K., & Sauerborn, J. (2006). Vascular plant species inventory of a Philippine lowland rain forest and its conservation value. *Biodiversity and Conservation, 15,* 1271–1301.
Lasco, R. D., & Pulhin, J. (1999). Forest land-use change in the Philippines and climate change mitigation. *Mitigation and Adaption to Climate Change Journal, 5,* 81–97.

Liu, D. S., Iverson, L. R., & Brown, S. (1993). Rates and patterns of deforestation in the Philippines: Application to geographic information system analysis. *Forest Ecology Management, 57,* 1–16.

Margraf, J., & Milan, P. P. (1996). Ecology of dipterocarp forests and its relevance for island rehabilitation in Leyte, Philippines. In A. Schulte & D. Schoene (Eds.), *Dipterocarp forest ecosystems* (pp. 124–154). Singapore: World Scientific.

Markewitz, D., Davidson, E., Moutinho, P., & Nepstad, D. (2004). Nutrient loss and redistribution after forest clearing on a highly weathered soil in Amazonia. *Ecological Applications, 14,* 177–199.

Masunaga, T., Kubota, D., Hotta, M., Shinmura, Y., & Wakatsuki, T. (1998). Distribution characteristics of mineral elements in trees of tropical rain forest, West Sumatra, Indonesia. In A. Schulte & D. Ruhiyat (Eds.), *Soils of the Tropical Forest Ecosystems* (pp. 168–174). Berlin: Springer- Verlag.

Müller-Edzards, C. (1996). *Development of sustainable and ecologically sound agroforestry systems in Leyte, Philippines* (p. 38). TOB Publication F-11/10e Tropenokelogisches Begleitprogram, GTZ, Eschborn, Germany.

Myers, N., Mittermeier, R. A., Mittermeier, C. G., da Fonseca, G. A. B., & Kent, J. (2000). Biodiversity hotspots for conservation priorities. *Nature, 403,* 853–858.

Navarrete, I. A., & Asio, V. B. (2014). Research productivity in soil science in the Philippines. *Scientometrics, 100,* 261–272.

Navarrete, I. A., & Tsutsuki, K. (2008). Land-use impact on soil carbon, nitrogen, neutral sugar composition and related chemical properties in a degraded Ultisol in Leyte, Philippines. *Soil Science and Plant Nutrition, 54,* 321–331.

Navarrete, I. A., Tsutsuki, K., & Asio, V. B. (2013). Characteristics and fertility constraints of some degraded soils in Leyte, Philippines. *Archives of Agronomy and Soil Science, 59,* 625–639.

Navarrete, I. A., Asio, V. B., Jahn, R., & Tsutsuki, K. (2007). Characteristics and genesis of two highly weathered soils in Samar, Philippines. *Australian Journal of Soil Research, 45,* 153–163.

Navarrete, I. A., Tsutsuki, K., Asio, V. B., Jahn, R., & Kondo, R. (2009). Characteristics and formation of rain forest soils derived from late Quaternary basaltic rocks in Leyte, Philippines. *Environmental Geology, 58,* 1257–1268.

Navarrete, I. A., Tsutsuki, K., Asio, V. B., Masayuki, T., & Sueta, J. (2011). Chemical, mineralogical and morphological characteristics of a late Quaternary sedimentary-rock derived soils in Leyte, Philippines. *Soil Science, 176,* 699–708.

Nussbaum, R., & Hoe, A. L. (1996). Rehabilitation of degraded sites in logged over forest using Dipterocarps. In A. Shulte & D. Schone (Eds.), *Dipterocarp forest ecosystems, towards sustainable management* (pp. 446–463). Singapore: World Scientific.

Ohta, S., & Effendi, S. (1992). Utisols of the lowland dipterocarp forest in East Kalimantan, Indonesia, II. Status of carbon, nitrogen and phosphorous. *Soil Science and Plant Nutrition, 38,* 207–216.

Otsuka, H., Briones, A. A., Daquiado, N. P., & Evangelio, F. A. (1988). *Characteristics and genesis of volcanic ash soils in the Philippines. Technical Bulletin.* Tropical Agriculture Research Center: Japan.

Pulhin, J. M., et al. (2006). Historical overview. In Chokkalingam U, Carandang AP, Pulhin JM, Lasco RD, Peras RJJ, Toma T (Eds.), *One century of forest rehabilitation in the Philippines: approaches, outcomes and lessons.* Bogor, Indonesia: Center for International Forestry Research (CIFOR).

Rebugio, L., Pulhin, J., Carandang, P., Peralta, E., Camacho, L. & Bantayan, N. (2007). *Forest restoration and rehabilitation in the Philippines*. Retrieved October 8, 2014 from http://www.iufro.org/download/file/7382/5120/Philippines_pdf/.

Richter, D. D., & Babbar, L. I. (1991). Soil diversity in the tropics. *Advances in Ecological Research, 21,* 315–389.

Roth, D. M. (1983). Philippine forests and forestry: 1565–1920. In R. P. Tucker and J. F. Richards (Eds.), *Global deforestation and the nineteenth century world economy* (pp. 30–49). Durham, N.C: Duke Press Policy Studies, Duke University Press.

Sanchez, P. A. (1976). *Properties and Management of Soils in the Tropics* (p. 618). New York: John Wiley & Sons.

Sanchez, P. A., & Logan, T. G. (1992). Myths and science about the chemistry and fertility of soils in the tropics. In *Myths and Science of Soils of the Tropics* (pp. 35–46). SSSA Special Publication No. 29, Madison, Wisconsin.

Schlichting, E., Blume, H. P., & Stahr, K. (1995). *Bodenkundliches practicum* [Soil Science Training] (2nd ed.) (p. 295). Berlin: Blackwell.

Schneider, T., Ashton, M. S., Montagnini, F., & Milan, P. P. (2014). Growth performance of sixty tree species in smallholder reforestation trials on Leyte, Philippines. *New Forests, 45,* 83–96.

Schulte, A., & Ruhiyat, D. (1996). *Soils of tropical forest ecosystems* (p. 206). Berlin: Springer-Verlag.

Simon, J. D., Natividad, N. M., Amaba, R. M., & Demen, T. P. (1975). *Soil survey of Samar provinces, Philippines*. Manila: Bureau of Print.

Schroeder, P. (1995). Organic matter cycling by tropical agroforestry systems: A review. *Journal of Tropical Forest Science, 7,* 462–474.

Schulz, J. P. (1960). *Ecological studies on rainforest in Northern Suriname* (p. 267). Uitgeveers Maatschappij, Amsterdam: North Holland.

Uehara, G. (1978). Mineralogy of the predominant soils in tropical and sub-tropical regions. In C. S. Andrew & E. J. Kamprath (Eds.), *Mineral nutrition of legume in tropical and subtropical soils*. CSIRO: Melbourne.

Van Raij, B., & Peech, M. (1976). Electro chemical properties of some Oxisols and Alfisols of the tropics. *Soil Science Society of America, Proceedings, 36,* 587–593.

Villavicencio, V. (1987). Philippines. In C. Lin Sien (Ed.). *Environmental management in Southeast Asia* (p. 77–107). Faculty of Science. National University of Singapore.

Wernstedt, W. H., & Spencer, J. E. (1967). *The Philippine island world: A physical, cultural and regional geography*. Berkely, LA: University of California Press.

Whitmore, T. C. (1998). *An introduction to tropical rainforests* (p. 282). Oxford: Oxford University Press.

World Bank. (2016). *The cost of fire: an economic analysis of Indonesia's 2015 fire crisis* (p. 10). Jakarta: World Bank Group.

Zech, W., & Drechsel, P. (1998). Nutrient disorder and nutrient management in fast growing plantations. In A. Schulte & D. Ruhiyat (Eds.), *Soils of Tropical Forest Ecosystems* (pp. 99–106). Berlin: Springer-Verlag.

Zikeli, S., Asio, V. B., & Jahn, R. (2000). Nutrient status of soil in the rain forest of Mt. Pangasugan, Leyte, Philippines. *Annals of Tropical Research, 22,* 78–88.

Zimmermann, M. H., & Brown, C. L. (1971). *Trees structure and function* (p. 336). New York: Springer-Verlag.

Chapter 6
Indonesian Peatland Functions: Initiated Peatland Restoration and Responsible Management of Peatland for the Benefit of Local Community, Case Study in Riau and West Kalimantan Provinces

Haris Gunawan

Abstract Indonesian peat swamp forest ecosystems provide various environmental services, which are important locally, nationally, and globally. However, due to increasing land scarcity, the pressure to utilize them for agriculture is increasing. This chapter focuses on a case study in Riau and West Kalimantan to discuss the socioeconomic and environmental values of peatland through examples of financial analysis and economic valuation and discussion based on fieldwork data. It shows that the opportunity cost of CO_2 emission reductions by conserving peat swamp forests from conversion to oil palm plantation ranged from USD $3.7 to 8.25/t CO_2e, which is far higher than the current registered emission reduction compensation price. Opportunity costs are higher than the carbon market price, and a carbon market is not available currently, especially for peat forest conservation. This chapter clarifies what models are viable taking the case study as a point of departure and calls for urgent development strategies to establish viable compensation alternatives to landholders beyond the carbon market. Peat conservation measures imply high opportunity costs, however indigenous and adaptive plants show economic promise to help further develop markets, paludiculture techniques, and management options to rewet peatlands.

Keywords Initiated peatland restoration · Peatland functions · Riau West Kalimantan · Socioeconomic costs

H. Gunawan (✉)
Peatland Restoration Agency of Republic of Indonesia, Jakarta, Indonesia
e-mail: haris.gunawan@brg.go.id

M. Lopez and J. Suryomenggolo (eds.), *Environmental Resources Use and Challenges in Contemporary Southeast Asia*, Asia in Transition 7,
https://doi.org/10.1007/978-981-10-8881-0_6

Introduction

Tropical peatlands are important for global, national, and local environmental functions and simultaneously, they have come to be appraised for their growing social and economic importance. Among various environmental functions they provide, the capacity to sequester and store high amounts of carbon (C) is the most prominent one. At a global scale, peatlands store an estimated 120 gigatonnes (Gt) of C or approximately 5% of all global terrestrial carbon (Rieley and Page 2005). The 14.9 million ha (Mha) of Indonesian peatlands (Ritung et al. 2011) store about 27 Gt C below ground in the form of peat organic carbon or an average of about 600 t C ha^{-1} m^{-1} (Agus et al. 2013a). In addition to peat soil carbon, peatlands also store carbon in plant biomass, which varies from about 2 t C ha^{-1} in annual crop systems to about 200 t C ha^{-1} in undisturbed forests (Agus et al. 2014a). Peatland carbon storage capacity is also important at a national scale because of high government reliance on peatland as one of the main contributors to a 26% C emission reduction pledge as stipulated in Presidential Regulation No. 61/2011. At a local scale, peat domes, resting on water-tables and fed by rainwater, have a crucial function in hydrological control. They can absorb water up to 13 times the mass of dry peat during the rainy season and release some of this water gradually during the dry season. This helps to mitigate the surrounding areas from floods and droughts (Agus and Subiksa 2008).

Peatlands in Indonesia are an important repository of unique biodiversity, prevent saline water intrusion from the sea into inland areas, and provide a cooling effect in surrounding areas due to high water storage (Parish et al. 2012). Most Indonesian peatlands remained undeveloped until the late 1960s, due to difficulty in accessing them (Kobayashi 2008). They also remained underutilized because of their inferior nature as productive agriculture land, mainly due to low fertility and water saturated conditions (Agus and Subiksa 2008). However, since the 1970s, peat swamp forests have been opened up by logging operations in Sumatra and Kalimantan (Kobayashi 2008). When peat is drained and fertilized, it can support the growth of almost all kinds of crops including vegetable, food, and plantation crops (Agus et al. 2012; 2013b), supporting the livelihood of local people and contributing to the national economy.

Peat swamp forests also possess a unique ecosystem function when they are pristine, but these are fragile and can be rapidly degraded through human activities (Baccini et al. 2012). Researchers have shown how land clearing and drainage network development contributes to their degradation (Page et al. 2011). There are wide potential impacts that may result from the development of peatlands for agriculture and these include soil subsidence, flooding, water shortage and pollution, fires and air pollution, habitat loss and biodiversity change, as well as changes in socioeconomic conditions. The pressure whether to conserve peatland in many cases is as strong as to develop it for agriculture, settlement, and mining purposes. The dilemma that Indonesia and other Southeast Asian nations that possess peat swamp forests face is the pressure to prioritize conservation, whilst not neglecting people's needs to use peatlands for development. Concurrently, time, uncontrolled use, and unwise peatland management practices do pose serious risks to local people, the public at large, and future generations. To respond to issues in Indonesia, this chapter discusses the socioeconomic and environmental values of peatlands and provides examples of financial analysis and economic valuation to conclude with suggestions for a way to balance environmental and economic needs.

Social Functions of Peatland: From Traditional to Intensive Use

Peatlands and people have been connected through intricate histories of cultural development in Indonesia's economy and the different intensity of peatland use is related to time, their vicinity to centers of development, access to capital, and domestic and global market demands on products that can be produced in them. Indonesia's peatland development is a relatively recent development, initially introduced for subsistence farming (Andriesse 1988) and then for plantation crops, annual crops, and multistrata systems. Their use can range widely and they can be a source of raw materials for industry and forestry under large-scale intensive cultivation or be used by smallholders for subsistence as well as commercial crop cultivation. Over the last 20 years, forest areas have decreased to make way for agriculture and other uses, the most important include plantations (oil palm, timber trees such as *Acacia*, rubber and various perennial tree crops), agroforestry, and annual crops (paddy system, vegetables, and cereal crops) (Gunarso et al. 2013). Sago forest plantations can also be found in parts of Papua such as the Mimika area and they require no drainage system or fertilization.

The shift to using peatland for agriculture production has significantly contributed to household, district, provincial, and national level income increases. Among agricultural uses, acreage allocated to oil palm plantation on peatland has increased rapidly from 4% in 2000 to 6% in 2005 to 10% in 2010 relative to the total Indonesian peatland area of 14.9 million ha (Agus et al. 2014a, b). This increase is attributed to the ease of market accessibility and the profitable use of land. Therefore, scholars have noted that abandoning agricultural practices and restoring the land to natural conditions would imply severe socioeconomic consequences (Van Beukering et al. 2008). However, there is no doubt that the uncontrolled use of peatland will degrade the benefits of peat forest environmental services.

Despite intense use in some areas, large areas of peatland have also been abandoned and covered by shrubland due to various reasons including unavailable labor and capital as well as remote locations from markets. Shrub-covered wastelands cover an estimated 22% of the total peatland area (Agus et al. 2014b) and tend to be the origin of many peat fires during the dry season (Agus et al. 2013a).

Subsistence Use of Peatland and Its Transformations

Until the late 1960s, people never lived permanently or cultivated crops regularly on peatlands (Rieley and Page 2005). Those who lived in rural areas depended on forests for a wide range of goods and services and subsequent forest conversion impacted on their livelihoods and culture. Research has clearly shown that when forests are replaced by oil palm monoculture, communities lose access to timber (for construction), rattan, and jungle rubber gardens (Sheil et al. 2009). Many of Indonesia's indigenous people practice shifting cultivation, whereas companies generally prefer hiring migrant workers with backgrounds in sedentary agriculture. This has led to ethnic conflicts between newcomers and indigenous groups (See Anderson 2013). Momose (2002), for example, has revealed that indigenous Malay villages in peat swamp forests along the Kampar River, in Riau Province, conducted agricultural activities such as shifting rice cultivation as well as fishing, hunting, non-timber forest product (NTFP) harvesting, and logging using their local knowledge. Malay people have contended that *ongka*, a traditional way of logging, is a sustainable form of peat swamp forest use. In Kalimantan, the indigenous Dayak people have long depended on natural wetlands, mainly consisting of peat swamps and freshwater swamps, for their livelihood and have engaged in shifting rice cultivation, the harvesting of forest products, and fisheries (Saman and Limin 1999). However, the Malay, Buginese, and Banjarese who live in estuary areas have historically had stronger relations to peatland and pioneered their sustainable use. The Chinese community in West Kalimantan, as well as Banjarese and Buginese in Sumatra, were also among the initiators of peatland use, who were followed later by Javanese and Balinese transmigrants. Researchers have also shown that native Mamuju farmers in West Sulawesi believe that peatland can be adapted for almost all kinds of crops, except for cocoa (Rina und Noorginayuwati 2012). Shallow peats that are generally fertile and suitable for agriculture have tended to be prioritized and their main uses have included swamp paddy systems, fishing, duck rearing, and water buffalo farming. Communities have historically adapted to seasonal water fluctuations and paddy farming has been practiced during the dry season while fishing and duck rearing have been practiced during the wet season. In this way, their management systems are perceived to function in a sustainable fashion. In other words, local farmers have a local understanding rooted in their interactions with their ecological environments and are clearly conscious of the fact that continuous draining and land exploitation will result in subsidence and a decline in soil fertility (Noorginayuwati et al. 2006).

Traditional systems are still maintained in some areas of peatland, although in the last couple of decades traditional farming and gathering have not been the sole sources of income. Increasingly, people have diversified their livelihood activities to work in both the service and commercial sectors as drivers, traders, laborers, and civil servants to increase income. Despite this diversification of income sources, nature-friendly forms of traditional farming have been maintained (Umar et al. 2014) and multiple cropping systems are more sustainable within the ecological conditions that exist (MacKinnon et al. 2000).

Furukawa (1994) has also described the different groups of people who make use of peat swamp forests in Riau province. One group is the Malay people who have been characterized as possessing multiple traditional farming activities that include fishing, hunting, gathering *jelutung* latex and rubber, rice and coconut farming, and nomadic forms of trading. Other immigrant groups such as the Buginese and Banjarese who have occupied the tidal areas of the eastern part of Sumatra, have built rice fields and became permanent residents there.

From the 1970s onwards, transmigrant farmers from densely populated areas of Java introduced more intensive forms of farming. This was partly due to relatively small parcels of land (2.25 ha per household (HH)) they were entitled to, compared to that of local farmers who used to have the privilege to clear and claim the forest and possess a larger share. Lowland rice crop systems were the mainstay among transmigrants and it is usually in rotation with *palawija* (secondary crops such as maize or beans) during the dry season (Noor 2001).

In general, local Banjarese, Dayak, Buginese, and Chinese systems are more diverse, combining annual and perennial crops, compared to transmigrant agriculture systems. Diverse systems such as those traditionally practiced by the Banjarese people have been gradually adopted by Javanese transmigrant farmers as shown by the succession of some of the lowland rice systems and perennial crops such as coconut and a combination of fruit trees with fish ponds in Purwosari, Tamban, and South Kalimantan (Collier 1982; Haris 2001). However, despite a relatively large area of land managed by local people, simple water transportation facilities limit access to markets and thus the management of large areas have been limited by investment and labor availability (Umar et al. 2014). Therefore, with time, traditional systems have incrementally diminished. Within this context, the development of peatland agriculture has adapted to the market and local socioeconomic and institutional capacity that supports economic activities. At present, intensive smallholder vegetable, fruit, and perennial crop farming can be found in peatland areas (Agus et al. 2012). A study by Van Beukering et al. (2008: p. 32) in Central Kalimantan revealed that several factors influence farmers' choice of crop varieties:

- Familiarity with and knowledge about the species. This is especially applicable for rubber as opposed to oil palm;
- Improvement in accessibility to seeds and nurseries;
- Improvement of accessibility to markets;
- Importance of the species for personal consumption (such as rice);
- Availability and size of local markets for alternative crops; and
- Frequency of harvest. Farmers' preference to rice is that they can harvest twice per year, ensuring a more frequent source of income (especially for farmers who have shallow peat). The same applies for rubber trees in terms of harvest frequency.

Van Beukering et al. (2008) further explained that two factors influenced land use strategies among Central Kalimantan farmers:

(a) **Type of peatlands**. Deep peatland tends to be used less intensively for rubber because of difficulty with drainage and low productivity. Shallow peat (less than 1 m) tends to be used more intensively for rice, vegetables, and fruits.
(b) **Plot size**. People with just one hectare of land, use their entire plot for food crops, especially rice, whereas owners of larger plots tend to have more diverse land uses.

Intensive Usage

The importance of peatland for agriculture has increased not only for local people but also for large-scale plantations. In recent years, large areas of tropical peatlands have been cleared and drained, especially for cash crops such as oil palm and timber plantations. Increasingly, to some extent, vegetable crops have undergone intense cultivation, especially in the vicinity of urban areas (Agus et al. 2012). Present conversion rates mean that approximately 51% of the total peatland area of 14.9 Mha remain as natural (primary and secondary) forests. 18% of peatlands have been used for various kinds of agriculture and forest plantations and 3% have been utilized for various uses such as building and mining.

The main areas with rapid development of oil palm plantation on peatland are Riau (the Indonesian province with the largest area of peatland), Jambi, Central Kalimantan, West Kalimantan, and East Kalimantan. Since 2016, the Indonesian government has enacted a policy to restore degraded peatland in seven priority provinces; one of them is West Kalimantan, with a total protected area of 28,136 ha, the permitted cultivation area is 64,078 ha, and nonpermitted area is 27,239 ha (See Fig. 6.1). In Riau province, the total targeted area for restoration is 815,180 ha, protected area is 43,810 ha, permitted cultivation area is 707,836 ha, and nonpermitted cultivated area is 43,810 ha (Fig. 6.2). Areas for timber (Acacia) plantations are mainly in Riau and Jambi Provinces (Gunarso et al. 2013). Under smallholders, intensive uses are for vegetable crops, fruit trees, and oil palm plantation. Areas of vegetables and fruits are relatively small because of high capital and labor requirements. The intensification of peatland use involves the development of drainage canals and use of chemical fertilizers and more often than not, land preparation involves fire. Most of the intensification is characterized by monoculture systems, which ultimately cause the loss of some of the peatland's primary ecosystem functions.

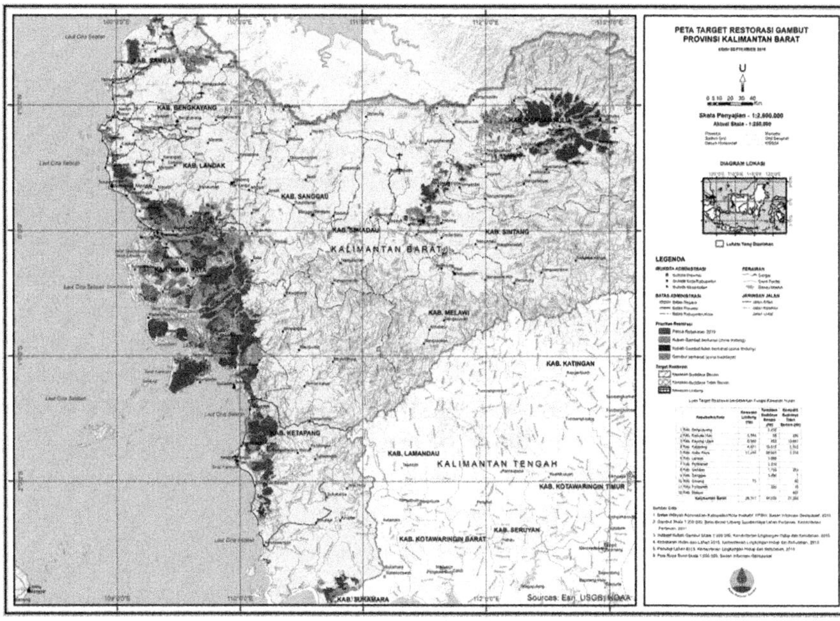

Fig. 6.1 Peatland restoration target area by peatland restoration Agency in West Kalimantan Province. *Source* Peatland restoration agency, Republic of Indonesia, 2017

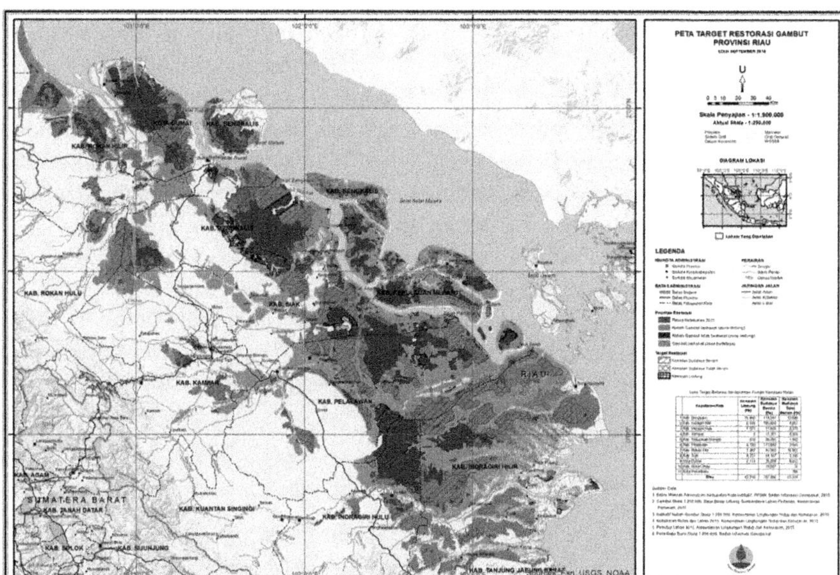

Fig. 6.2 Peatland restoration target area by peatland restoration agency in Riau Province. *Source* Peatland restoration agency, Republic of Indonesia, 2017

Multifunctionality of Peatland

The various tangible and intangible functions of peatlands are generally poorly understood or incompletely recognized and it is often the case that different stakeholders consider peatland from their own perspectives. This is one of the fundamental root causes of uncoordinated use, degradation of peatland, and repeated conflicts among stakeholders. In general, in its original form, peatland provides various ecosystem services, some of which are intangible to the public because their values are not marketable. Intangible services include (1) water retardation and thus regulation for landscape hydrology function, (2) carbon sink and thus carbon storage, and (3) a diverse ecological niche for various flora and fauna and thus biodiversity function (Agus and Subiksa 2008). With an increase in the population and higher demands on land, the role of peatland has become more important for employment, food security, and income generation.

Peatland Ecosystem Functions

Ecosystem functions are the free services that ecosystems provide to humanity (WRI 2005) and their services include the goods provided by the ecosystems and the different functions they perform, both of which directly or indirectly benefit human well-being and that of other species, flora and fauna. Peat swamp forests play an integral role in maintaining a wider regional ecosystem balance and critically support social and economic systems through the functions they provide. Table 6.1 presents an overview of the main peatland ecosystem functions, categorized into direct uses, indirect uses, and nonuse values. These three types of values form the total economic value of peatlands and constitute the beneficial outcome of the hydrological, chemical, and biological processes within the ecosystem. In particular, hydrological functions are especially important on a local scale. Peat is hydrophilic, absorbs water during the rainy season, and releases it slowly during dry periods. Intact peatlands possess a great potential to mitigate the loss of life and damage to infrastructure by reducing flooding downstream of peatlands. Intact peatlands can also help to maintain minimum flows in rivers during the dry season and can minimize drought stress in peat areas surrounding peat domes. In addition, the wetland ecosystem of peatlands helps to prevent saline water intrusion from the sea into inland areas.

Table 6.1 Ecosystem services provided by peat swamps ecosystem

Peatland values	
Direct use *(production functions)*	•Storage of water •Ecotourism
Direct extractive use of biodiversity	•Food (e.g. Fish) •Medicinal plants •Ornamental plants •Aquarium fish •Timber •Non-timber forest produce, such as rattan and other plants for construction purposes, •Fuel and handicrafts
Indirect use *(regulation functions)*	•Storage and sequestration of carbon •Reduction of downstream flood by its high (up to 13 times its dry mass) water absorbing capacity •Reduction of drought impact by maintenance of base (minimum) flows in rivers by releasing water slowly during dry periods •Prevention of saline water intrusion
Non use	•Spiritual, historical, and cultural values
	•Aesthetic values
Biodiversity attributes	•e.g., species richness and endemism

Source ASEAN (2006)

Integrated Functions of Peatlands

Land use change is often accompanied by changes to people's livelihood settings and the environmental functions of peat. Deterioration of peatlands is usually associated with extensive clearing and draining that, in turn, results in significant socioeconomic changes and the loss of environmental quality. These losses can sometimes lead to tensions between key stakeholders at local, regional and international levels. The Mega Rice Project (MRP) in Central Kalimantan, launched by the former President Soeharto on December 26, 1995, through Presidential Decree No. 82/1995, with the objective of developing approximately one million hectares of peatland for food crop production (especially rice), is an example of mass peatland development that ended in mass deterioration. The MRP, covering an area of 1,457,100 ha, left many problems that included;

1. Socioeconomic: Nonpayment of land compensation fees and loss of means of daily subsistence by the local inhabitants as a result of damage to their traditional fishponds (locally called *beje*) and other forest products such as rattan, latex, food, and medicines (Jaya 2004).
2. Biophysical: A decrease in water quality from increased acidity and the lowering of the water table leading to a decrease in water regulation functions of the land and loss of vegetation cover.

The MRP implementation was reevaluated in May 1998 and this revealed that from the outset, the project had been misguided in terms of its rationale, strategy, planning, implementation, monitoring, and budget allocation. Research has subsequently shown that the mass transformation on the peatlands has led to a rise in frequent fires related to droughts (Putra et al. 2008). A review of the MRP also concluded that from the 40-50% of the total project area that consisted of peatlands, only 400,000 ha of the peatland were actually suitable for agriculture, and the rest should have been conserved as an environmental safeguard.

Another example of negative socioeconomic impacts is the expansion of palm oil plantations on forest-dependent communities. Many people who live in rural areas depend on forests for a wide range of goods and services. The conversion of forests has impacted considerably on the livelihoods and culture of indigenous populations and the environmental problems associated with intensive agricultural development of peatland have included artificial drainage inducing subsidence of the ground surface; hydrological changes that may eventually lead to the loss of the peat swamp environment along with its ecological functions (e.g., biodiversity, water storage, hydrological regulation, and carbon storage). Furthermore, fertilizers and pesticides led to long-term water pollution and the burning of peat and vegetation during land clearance resulting in intensive air pollution (Rieley and Page 2005; Parish et al. 2012).

Due to the fragility of peat swamp ecosystems and the high environmental services they provide, it is essential to use them in a wise and sustainable way. Joosten and Clarke note that "the wise use of mires and peatlands as those uses of mires and peatlands for which reasonable people now and in the future will attribute no blame. The word 'use' is employed in its widest meaning, including conservation and nonuse" (Joosten and Clarke 2002: p. 19). Building upon this definition, Rieley and Page (2005) have stated that the wise use of peatlands can maintain their functions and values to support environmentally compatible development for sustainable agriculture. They thus suggest a multiple wise use of tropical peatlands that optimizes their economic, social, and ecological values. This is done by harvesting renewable resources in a sustainable way while conserving nonrenewable resources and maintaining the attributes and functions of the land.

This approach implies that limited areas of peatland can be converted to support small-scale agriculture. In such cases the following guidelines should be considered: the selection of the areas should ensure minimal impact on the ecosystem's functions; the suitability of the crop for peatland cultivation should be determined carefully; and finally, drainage networks should be carefully constructed and be kept shallow for agricultural use.

The Economics of Peatlands: Development Versus Conservation Uses

Peatlands produce marketable and nonmarketable products. Marketable products are those related to agricultural production such as palm oil, latex, corn, rice, fruits, and vegetables. In some, but rare instances, the peat itself is harvested and used as fuel or growth media in nurseries and thus it is marketable. On the other hand, peatland under natural forest conditions provides several ecosystem services (Agus and Subiksa 2008; Parish et al. 2012) which can be valued in economic terms, but under the current economic system, the market seems not yet consider them as something valuable (Goda et al. 2006). These services may include those listed in Table 6.1.

Market Values of Peatlands

Herman et al. (2009) conducted an economic analysis of oil palm plantations on peat and mineral lands and concluded that oil palm plantation development was as economically feasible on peatlands as on mineral lands. Their study assumed a range of crude palm oil (CPO) prices of IDR 6000 (US\$0.45) to 10,000 (US\$ 0.75) kg^{-1} and palm kernel prices of IDR 4500 (US\$0.35) to 6500 (US\$0.50) kg^{-1}. Under those assumptions oil palm plantations on mineral land generated a net present value (NPV) of IDR 0.90 million (US\$67.5) to 3.01 million (US\$225) ha^{-1} yr^{-1} under the nucleus estate scheme and IDR 0.70 million (US\$52.5) to 2.53 million (US\$189.8) ha^{-1} yr^{-1} under large plantation scheme. On peatland, the NPV was IDR 0.74 million (US\$57.8) to 2.85 million (US\$213.8) ha^{-1} yr^{-1} under the nucleus estate scheme and IDR 0.55 million (US\$41.3) to 2.39 million (US\$179.3) ha^{-1} yr^{-1} under large plantation scheme. Other indicators of financial analysis, the benefit/cost (B/C) ratio and the internal rate of return (IRR) also support the argument of economic feasibility. With the assumption of CPO price of IDR 10,000 kg^{-1} (US\$0.75) and palm kernel prices of IDR 6500 kg^{-1} (US\$0.50), under the nucleus estate scheme, the IRR ranged from 33 to 34% (which are well above the discount factor of 15%) and the B/C ratio ranged from 1.4 to 2.3 on mineral land. On peatland, they ranged from 31 to 33% and 1.3 to 2.2, respectively.

In addition to palm oil, Herman et al. (2009) have conducted research on various other crops including vegetables, pineapple, maize, and rubber: adaptable and competitive on peatland. The results of their financial analysis in West Kalimantan, based on different scenarios of prices in 2009, are presented in Table 6.3. Oil palm was shown to be superior to rubber and comparable to intensive cultivation pineapple system in terms of the NPV. However, in many cases, smallholder farmers have shown a preference for rubber over oil palm and tradition pineapple over the more intensive system. In the case of rubber, low input and continued harvest are important considerations and in the case of traditional pineapple, low input is the main factor because many farmers cannot afford the high input system. For annual crops, vegetable crops provide the highest profits, but capital and labor involved in vegetable crops were also high. Maize required lower capital for supplies and labor with much lower profits compared to vegetables. Vegetables and pineapple were also subject to high price fluctuations. What is important to note is that capital for oil palm and rubber could be an impediment factor. Smallholder farmers tended to gradually develop their farms to avoid high and abrupt investment at the beginning of farm development. Despite very high net profits from vegetable farming on peatland (Table 6.2), it did not change rapidly. Instead, *sawah* (paddy field in rotation with secondary crops of either maize or vegetable crops) was the form that developed most rapidly. Oil palm on peatland commenced in the study districts in West Kalimantan and its expansion is expected to be steady (shown in Fig. 6.3) or possibly show an acceleration in cultivation. Areas of traditional rubber plantation were decreasing and this seemed to be related to the higher incentives of other commodity developments.

Table 6.2 Analysis of capital, net profits (for annual crops) and the net present values, NPV (for perennial crops)

Crop	Capital (IDR ha^{-1})	Net profit (IDR ha^{-1} yr^{-1})	Annualized NPV at 15% discount factor (IDR ha^{-1} yr^{-1})		
			Scenario 1 (S1)	Scenario 2 (S2)	Scenario 3 (S3)
Oil palm	30,437,000	–	822,248	1,769,473	2,716,698
Rubber	15,230,000	–	400,080	1,108,869	1,817,658
Pineapple, traditional	3,827,500	–	242,421	529,330	816,240
Pineapple, intensive	9,947,500	–	1,184,833	1,864,897	2,544,960
Vegetables	2,500,000	10,009,000	–	–	–
Maize	1,250,000	2,226,102	–	–	–

Remarks: Crude palm oil price for Scenario 1 (S1) in IDR kg^{-1} = 6000; S2 = 8000; S3 = 10000. Palm Kernel price for S1 = 4000; S2 = 4500; S3 = 5000. Crumb rubber prices in IDR kg^{-1} for S1 = 7000; S2 = 11,000; S3 = 15,000. Pineapple fruit for S1 = 400; S2 = 500; S3 = 600.
Source Herman et al. (2009)

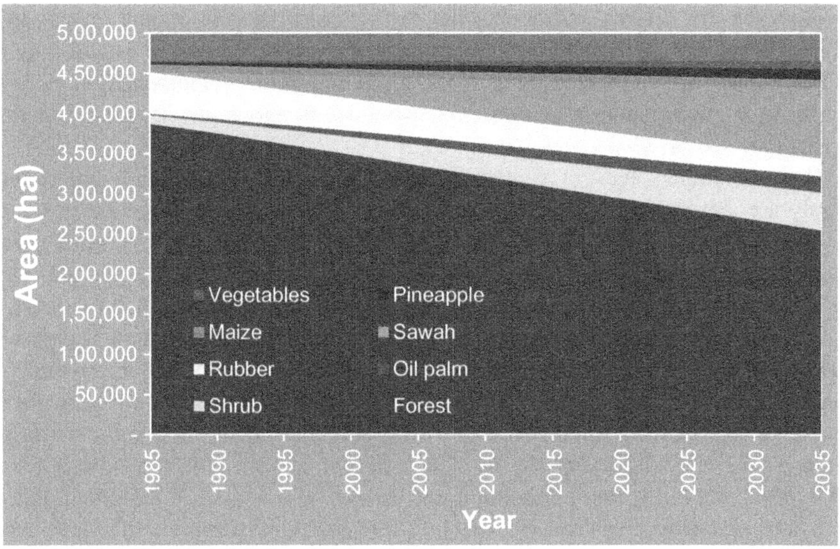

Fig. 6.3 Historical (1985–2009) and linear projection until 2035 of land use change in Kubu Raya and Pontianak Districts, West Kalimantan Province. *Source* Agus et al. 2012

NonMarket Values of Ecosystem Functions

Economic valuation is an attempt to assign quantitative values to the goods and services provided by environmental resources, whether or not the goods are marketable. Many efforts by economists and others have been made in the last few decades to "internalize the externalities" by modifying market valuation and by placing more emphasis on ecosystems functions (Starrett 2000).

There exist a series of methods to calculate the economic valuation of peatlands according to the Ministry of the Environment Regulation No. 14/2012; (1) Total economic value of Natural Resources and Environment, and (2) Degraded Environment Economic Values. Using these methods, natural peat swamp forest ecosystems in Zamrud Nature Reserve in Siak district (Riau province, Sumatra island, 31,480 ha), and Sebangau National Park in Katingan regency (Central Kalimantan province, Borneo island, 568,700 ha) were valued at around IDR 2.95 trillion (US$221,338,500) and IDR 2.3 billion (US$172,569), respectively. The loss of environmental resources is an economic problem because the natural assets are lost or degraded, some perhaps irreversibly.

Any options for environmental resource management, either to leave it in its natural state, allow it to degrade, or convert it to another use, has implications in terms of the gains or losses of these valuable assets. The decision of what use to pursue for a given environmental resource, and ultimately whether the current rates of resource loss are "excessive," can only be made if these gains and losses are properly analyzed and evaluated. For example, preserving an area in its natural state involves the direct costs of preservation and this may include paying guards and rangers to protect and maintain the area and perhaps also the cost of establishing a "buffer zone" for the surrounding local communities. Development options are sacrificed if preservation is chosen, which implies foregone development benefits (opportunity costs).

The economic values of peatlands' functions of flood mitigation and water supply are substantial. Whiteman and Fraser (1997) estimated the value of these functions of about US\$91.60 (ha^{-1} yr^{-1}) or IDR 925,253 (ha^{-1} yr^{-1}). Furthermore, drained peatlands are very susceptible to destructive fires during dry periods. The 1997 fires that burned 2–3% of the land area of Indonesia had an economic impact of at least US\$9 billion or equivalent to IDR 90,909 billion (Van Eijk and Leenman 2004). These fires lead to adverse economic impacts through the destruction of commercial timber, plantations, and farmland; a reduction in tourism; the temporary shutdown of commerce, industry, and travel; and an increase in health care costs (Sastry 2002). In addition to the fires' destruction, smoke, and haze impair photosynthesis, thus lowering agricultural and forestry production in unburned areas. Fires also eliminate seeds and seedlings, further degrade hydrologic functions, and cause soil erosion (Schrier-Uijl et al. 2013; Tacconi 2003). Repeated fire events lead to soil subsidence and the risk of flooding, which destroys crops and increases carbon emissions. A cost-based study by Tacconi (2003) estimated that the fires of 1997 cost around US\$4.5 billion through losses in various sectors such as timber, tourism, transportation, and agriculture, in addition to the actual costs of fighting the fires. Furthermore, in 2014 the fires that occurred in the Riau peatlands caused economic losses of around IDR 15 trillion (US\$1,125,450,000) (BNPB 2014). It is important to note that early calculations of the total economic costs of the fires in 2015 in Indonesia alone exceeded US \$16 billion. This is more than double the damage and losses from the 2004 tsunami (which affected provinces in Indonesia and other countries), and equal to about 1.8% of Indonesia's GDP[1].

One approach to maintaining the environmental functions of peat forest is through implementing a sustainable harvesting system. Although shifting to a sustainable harvesting system reduces the net benefits of timber harvesting, one case study has suggested that this was more than offset by the increased nonmarket benefits, primarily hydrological, and carbon storage values (Wetlands International 2007). When the carbon price was relatively high, some experts have suggested to concentrate on the value of carbon captured and stored by peat swamp ecosystems. Preserving forests and peat swamps, which would otherwise be converted, and collecting the resulting recurrent revenues provided by the carbon offset market may be more lucrative for landholders in some areas than conversion to oil palm (Butler 2007) provided that such markets are in place.

---

[1]See Indonesia's fire and haze crisis. http://www.worldbank.org/en/news/feature/2015/12/01/indonesias-fire-and-haze-crisis Accessed 14 January, 2017.

Many economic valuation studies try to compare various scenarios and usually come to the conclusion that the sustainable use of natural resources creates much more value than their unsustainable use. Such results are often dependent on the inclusion of nonmarket ecological services because in the long term, the benefits of those are often higher than the marketed benefits. Yet in reality, forest conversion continues because of a lack of awareness of the wider (nonmarket) economic, social, ecological, and environmental benefits (Joosten and Clarke 2002) and because of the absence of incentives provided to landholders. Therefore, considering the value of carbon storage and other benefits provided by peatlands, policies need to be developed for compensation of the opportunity cost to landholders for providing public goods. This means that payments for environmental services in various forms must be realized.

Tradeoffs Between the Environmental and Development Objectives

Several indirect substitutions cost methods have been used to evaluate the economic values of ecosystem functions. These include the Hedonic Price Method (HPM), Contingent Valuation Method (CVM), Willing to Pay Method (WTP), and Replacement Cost Method (RCM) (Goda et al. 2006). HPM may not be suitable for peatland because it is about dealing with the amenities of land for housing environment that are reflected in land prices and wages. The CVM has been increasingly used for valuation of multifunctionality of agriculture. In this method, questionnaires are sent to the general public who benefit from the environmental functions and ask them how much they are willing to pay (WTP) for maintenance of such functions (Agus 2006). For example, if peat forest is important for water storage and water redistribution, the respondents (the local beneficiaries) are asked how much they are willing to pay for the forest conservation. RCM evaluates the cost incurred by individuals for restoring certain environmental quality. For example, for reducing heat among the people who live near, where peat forest used to be, they have to use electric fans or air conditioners to restore cooler conditions provided by the forest. The costs for purchasing and operating those appliances are called RCM.

For CO_2 emissions reduction, the approach of opportunity cost has been used widely (e.g., Gregersen et al. 2010; Herman et al. 2009; White and Minang 2011). In this case, opportunity cost can be defined as the ratio of change in profits (net present value, NPV) divided by the change in carbon stock. In other words, opportunity cost is NPV divided by the unit weight of GHG emissions as illustrated in Fig. 6.4. In Fig. 6.4, the change of a hectare of forest with the carbon stock of 250 t C ha^{-1} to a hectare of agricultural land with the carbon stock of 5 t C ha^{-1} causes the net change of 245 t C ha^{-1}. The estimated profits from agriculture are US $400 ha^{-1}, while forest profits are US $50 ha^{-1} and both are expressed in the NPV terms. So the difference in NPV is US $350 ha^{-1}. The opportunity cost of

Fig. 6.4 Illustration of carbon loss and profit gain from converting forest to agriculture *Source* White and Minang 2011

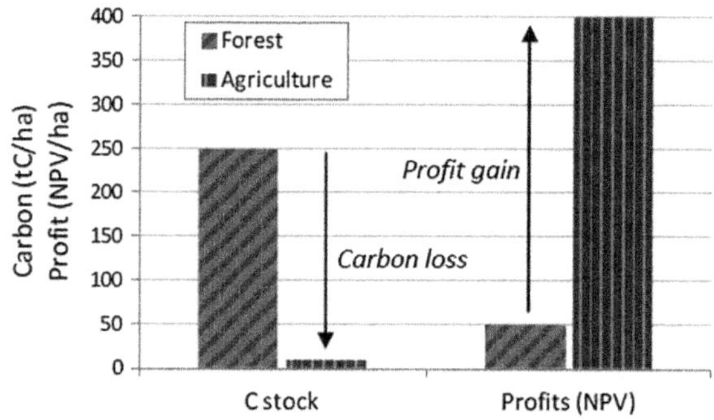

Table 6.3 Opportunity costs of conserving peat forest from conversion to oil palm plantation under the nucleus estate and large plantation schemes with the assumption that mean greenhouse gas annual emission under oil palm plantation is 64 t CO_2 (ha. yr)$^{-1}$ and under forest is zero (adapted from Herman et al. 2009)

Plantation model/ Stakeholder	Net present value (at 15% df)		Opportunity cost	
	IDR (ha yr)$^{-1}$	USD (ha yr)$^{-1}$	IDR/t CO_2e	USD/t CO_2e
Nucleus estate (NE)	2,845,000	310	44,520	4.84
Plasma farmer	*2,180,000*	*237*	34,034	*3.70*
Nucleus	*4,856,000*	*528*	75,886	*8.25*
Large plantation	2,387,000	260	37,253	4.05

Opportunity cost = NPV/64 t CO_2 (ha yr)$^{-1}$
Assumptions: Crude palm oil price was IDR 10,000 kg^{-1}; palm kernel price was IDR 6500 kg^{-1}

conserving or not changing forest to agriculture equals US $350 ha^{-1} divided by the 245 t C ha^{-1} not emitting C or equals US $1.43/t C. However, carbon emission compensation is based on carbon dioxide equivalents (CO_2e). A conversion factor of 3.67 is needed to translate t C to t CO_2e. So, the potential emissions of land use change is 245 t C ha^{-1} * 3.67 t CO2e/t C = 899 t CO_2e ha^{-1}. Therefore, the equivalent opportunity cost of avoided emissions is US $0.39/t CO_2e. From the perspective of the beneficiaries the expression in terms of per ton CO_2e is more popular, but for landholders, the more relevant expression is per ha. Thus, the opportunity costs of not converting the forest to agriculture, from the landholders' perspective is US $350 (ha yr)$^{-1}$.

Table 6.3 gives an example of the opportunity cost calculation of conserving Indonesian peat forests. The main driver of forest conversion in this example is oil palm plantation and so the NPV of peat forest versus oil palm plantation and annual average CO_2 emissions (from a 25-year cycle) of peatland under oil palm plantation were the main input data. The assumption of the mean annual CO_2 emissions under oil palm plantation of 64 t CO_2 (ha yr)$^{-1}$ include emissions from biomass loss from deforestation and peat decomposition.

This calculation shows that the opportunity cost range is from US \$3.7 to 8.25/t CO_2e or equivalent to the NPV of US \$237 to \$528 $(\text{ha yr})^{-1}$. Landholders expect a compensation is equivalent with the NPV. For carbon buyers, the opportunity cost of US \$3.7 to 8.25/t CO_2e indicates how much money they have to spend on compensation to reduce certain amount of emissions.

Scoping Activities in Peatland Restoration for Local Communities

Thousands of villages' peatlands are currently scheduled for restoration between 2016–2020 and communities will play a strong role in restoration activities. Three primary methods will be employed in restoring degraded peatland: rewetting, revegetation, and the revitalization of livelihoods (3R Integrated Method). Scoping activities in peatland restoration for local communities includes the following three areas:

1. Revitalization of Integrated Water Management to Speed up the Restoration of Peatland Ecosystems
During the dry season, large-scale companies tend to stem peat water in order to flood canals to prevent fires so that peatland become dry, plants die, and easily catch fire. In this case, there is an urgent need to research the hydrology of peat swamps in order to prevent disruption by fire. This may include activities including the trial creation of partition canals, water ropes, ponds, and the optimal normalization of canals so that they can simultaneously support the recovery process of ecosystem peatlands that burnt in the areas occupied by peat swamp communities. Partnership opportunities between large-scale companies and local communities in the improvement of water flow are needed to accelerate the recovery process of the peat swamp ecosystem. Studies should suggest water sharing interventions. The aim of restoration work is to improve hydrological regimes thus groundwater levels should be maintained at 0.4 m from the peatland surface. There is a need to develop groundwater level monitoring in real time to provide up to date information on the hydrological conditions. Peatland should be managed as a total hydrological unit at a landscape level.

2. Revegetation of Local Species for Building Productive Plantation Forests, and Facilitating Access to Timber and NTFPs Market in Order to Improve Community Welfare
Revegetation of degraded peat forests, in general, are well-known, but the development of agroforestry on peatlands by selecting the type of plant that is adaptive, without drying peat, and mixing with other types of plants (commercial timber, fruit trees, root crops, medicinal plant, and firebreaks plant) still requires further research. This is needed to develop productive plantation forests that can provide daily, short-, medium-, and long-term income for farmers, as well as prevent

disruption by fire. Identifying types of products that can become a leverage for community development can open up opportunities for policy intervention in order to improve local communities' welfare. There are a number of potential crops such as sago palm *(Metroxylon sago)*, coffee trees, coconut *(Cocos nucifera)*, jungle rubber *(Hevea sp)*, pineapple (*Ananas* sp) and many others forest tree species proven to grow in degraded peatland such as bintangur *(Callophylum sumtaranum)*, jelutung *(Dyera lowii)* and meranti rawa (*Shorea uliginosa*). These can develop into integrated farming systems, silvofishery, and agroforestry, locally known as kebun kayu campuran (mixed trees gardens).

3. Social Transformation to Strengthen Community Perspectives in Relation to Functions and Economic, Social and Environmental Benefits of Peatland
Communities living in peat swamp ecosystems can be categorized as both indigenous and immigrant communities. Indigenous communities are groups of people that used to live in harmony with their natural surroundings, but these have changed due to external influences. Immigrant communities, on the other hand, tend to be extractive. Fires have resulted in a loss of community livelihood for those who depend on peat swamps such as farming, livestock breeding, hunting, fishing, and peat swamp timber production. To support the improvement of peat swamp conditions, it is crucial to research those important values in society that need to be strengthened, especially those related to the alignment of economic, social, and environmental functions and benefits so that social transformation can occur effectively. There are a number of local communities that are enthusiastic to be involved in restoration activities, as illustrated by those in the village areas of Tebing Tinggi Timur and Bukit Batu (sub district in Riau Province). However, to guarantee the continuity of restoration activities, it is important to understand their basic necessities. Any meaningful restoration must provide benefits as a means for improvement of livelihood and strengthening the capacities for organization at a village level to weather change and move beyond current economic practices.

Conclusion

This chapter has discussed the socioeconomic and environmental values of peatland and included examples of financial analysis and economic valuation and discussion of fieldwork data in Riau and West Kalimantan, Indonesia. Despite the number of challenges and problems, especially with the responsible management of peatland, it does offer various kinds of socioeconomic functions that can potentially benefit local communities. Beyond the market and nonmarket values of peatland, there are also tradeoffs between environmental and development objectives. In the current global market's need for cheaper resources, there is a strong push for conversion to oil palm plantation in many parts of Indonesia, especially in peatland areas. The opportunity cost of CO_2 emissions reduction by conserving peat swamp forest from conversion to oil palm plantation ranged from US $3.7 to 8.25/t CO_2e. This is far higher than the current registered emission reduction compensation price. The opportunity cost is higher than the carbon market price, and a carbon market is not available currently, especially for peat forest conservation. Therefore, this chapter can offer lessons to policy makers in the government's ministries to consider the issue of socioeconomic functions of peatland, establish compensation alternatives to peatland-holders beyond the carbon market, and balance the needs between environmental protection and local community's development. This chapter suggests that peatland restoration requires intervention through three integrated methods: rewetting, revegetation, and the revitalization of livelihood. Local communities play a fundamental role in the success of peatland restoration achievement, as they have direct interaction with peatland areas. Therefore, transformation must involve the community in any peatland restoration activities and empower them in responsible management activities for improving the basis of their livelihood and ensuring the protection of swamp peatlands. Peat conservation measures imply high opportunity costs, however indigenous and adaptive plants show economic promise to further develop markets, paludiculture techniques, and change management options to rewet peatlands. Restoration measures by rewetting, revegetation, and revitalisation of livelihood can offer strong benefits to local communities and be potentially more sustainable than current drainage-base agriculture, but only if the environmental effects are made more explicit.

References

Agus, F. (Ed.). (2006). *Final Report: Country case studies on multifunctionality of agriculture in ASEAN Countries*. Jakarta: The ASEAN Secretariat, Japanese Ministry of Agriculture, Forestry and Fisheries (MAFF).

Agus, F., & Subiksa, I. G. M. (2008). *Lahan gambut: Potensi untuk pertanian dan aspek lingkungan.*[Peatland: farming potentials and environmental aspects]. Booklet. Bogor: Balai Penelitian Tanah dan World Agroforestry Centre (ICRAF) SE Asia. (in Indonesian).

Agus, F., Dariah, A., & Jamil, A. (2013b). *Kontroversi pengembangan perkebunan kelapa sawit di lahan gambut* [Controversies on conversion of peatland for palm oil plantation]. In H. Soeparno, E. Pasandaran, M. Syarwani, A. Dariah, S. M. Pasaribu and N. S. Saad (Eds.) *Politik Pembangunan Pertanian Menghadapi Perubahan Iklim* [Development politics of agriculture in the face of climate change], (pp. 452–473) (in Indonesian). Jakarta: IAARD Press.

Agus, F., Gunarso, P., & Wahyunto. (2014b). *Dinamika penggunaan lahan gambut* [The dynamics of peatland use]. In F. Agus, M. Anda, A. Jamil, & Masganti (Eds.), *Lahan Gambut Indonesia: pembentukan, degradasi dan potensi mendukung ketahanan pangan dan kualitas lingkungan* [Indonesian peatland: Formation, degradation, and potentials to support food security and environmental qualities] (pp. 85–100). Jakarta: IAARD Press. (in Indonesian).

Agus, F., Henson, I. E., Sahardjo, B. H., Harris, N., van Noordwijk, M., & dan Killeen., T. J. (2013a). *Review of emission factors for assessment of CO_2 emissions from land use change to oil palm in Southeast Asia*. Kuala Lumpur: Roundtable on Sustainable Palm Oil.

Agus, F., Santosa, I., Dewi, S., Setyanto, P., Thamrin, S., Wulan, Y. C., & Suryaningrum, F. (Eds.). (2014a). *Pedoman Teknis Penghitungan Baseline Emisi dan Serapan Gas Rumah Kaca Sektor Berbasis Lahan: Buku I - Landasan Ilmiah* [Technical guidelines of land-based emission and sequestration calculation: Book I - The Scientific Background]. Jakarta: Badan

Perencanaan Pembangunan Nasional, Republik Indonesia (Indonesian National Development Planning Agency, BAPPENAS). (in Indonesian).

Agus, F., Wahyunto, A., Dariah, E., Runtunuwu, E. S., & Supriatna, W. (2012). Emission reduction options for peatlands in the Kubu Raya and Pontianak Districts, West Kalimantan, Indonesia. *Journal of Oil Palm Research, 24,* 1378–1387.

Anderson, P. (2013). Free, Prior, and informed consent? Indigenous peoples and the palm oil boom in Indonesia. In O. Pye & J. Bhattacharya (Eds.), *The palm oil controversy in Southeast Asia: A transnational perspective*. (pp. 244–258) Singapore: ISEAS.

Andriesse, J. P. (1988). *Nature and management of tropical peat soils (Soils Bulletin 59)*. Rome: FAO- Food and Agriculture Organization of the United Nations.

ASEAN (Association of Southeast Asian Nations). (2006). *Rehabilitation and sustainable use of peatlands in South East Asia. Full project brief for international fund for agricultural development and global environment facility*. Jakarta: ASEAN Secretariat.

Baccini, A., Goetz, S. J., Walker, W. S., Laporte, N. T., Sun, M., Sulla-Menashe, D., Hackler, J., Beck, P. S. A., Dubayah, R., Friedl, M. A., Samanta, S., & Houghton, R. A. (2012). Estimated carbon dioxide emissions from tropical deforestation improved by carbon-density maps. *Nature Climate Change 2,* 182–185. doi:https://doi.org/10.1038/nclimate1354, ISSN: 1758-678X.

Beukering, P. van., Schaafsma, M., Davies, O., & Oskolokaite, I. (2008). *The economic value of peatland resources within the central Kalimantan peatland Project in Indonesia: Perceptions of local communities. Central Kalimantan Peatlands Project*. (Report E-08/05, June 28, 2008).

BNPB (Badan Nasional Penanggulangan Bencana). (2014). *Upaya Penanggulangan Bencana di Indonesia. Bahan Presentasi Kuliah Umum* [Efforts of disaster prevention in Indonesia. Presentation material for general lecture]. Riau Bebas Asap. Presentation for public lecture. Pekanbaru, 28 April 2014 (in Indonesian).

Butler, R. (2007). The Impact of Oil Palm Borneo. Retrieved July 10, 2016, from http://data.mongabay.com/borneo/borneo_oil_palm.html.

Collier, W. L. (1982). *Lima puluh tahun transmigrasi spontan dan transmigrasi pemerintah di tanah rawa Kalimantan* [Fifty years of spontaneous migration and government-sponsored migration in Kalimantan peatland]. In J. Hardjono (Ed.), *Transmigrasi: Dari Kolonisasi Sampai Swakarsa* [Transmigrasi*: From colonial times to self-reliance*]. Jakarta: Gramedia (in indonesian).

Eijk, P. van., & Leenman, P. (2004). *Regeneration of fire degraded peatswamp forest in Berbak national park and implementation in replanting programmes. Water for food and ecosystem project on: "Promoting the river basin and ecosystem approach for sustainable management of SE Asian lowland peatswamp forests." Case study Air Hitam Laut river basin Jambi province, Indonesia.* The Netherlands: Alterra & Green World Research, Wageningen.

Furukawa, H. (1994). *Coastal wetlands of Indonesia: Environment, subsistence and exploitation*. Kyoto: Kyoto University Press. Retrieved July 15, 2016.

Goda, M., Kada, R., & Yabe, M. (2006). Synthesis and conclusions. In F. Agus (Ed.), *Final report country case studies on multifunctionality of agriculture in ASEAN countries* (pp. 415–431). Jakarta: The ASEAN Secretariat & Japanese Ministry of Agriculture, Forestry and Fisheries (MAFF).

Gregersen, H., El Lakany, H., Karsenty, A., & White, A. (2010). *Does the opportunity cost approach indicates the real cost of REDD + ? Rights and realities of paying for REDD+*. Washington, D.C: CIRAD & Rights and Resources Initiative.

Gunarso, P., Hartoyo, M. E., Agus dan, F., & Killeen, T. J. (2013). *Oil palm and land use change in Indonesia, Malaysia and Papua New Guinea*. Kuala Lumpur: Roundtable on Sustainable Palm Oil.

Haris, A. (2001). *Manajemen Lahan Orang Banjar* [The land management of the Banjarese]. Fakultas Pertanian Universitas Lambung Mangkurat, Banjarbaru (Unpublised manuscript; in Indonesian).

Herman, Agus, F., & Las, I. (2009). *Analisis finansial dan keuntungan yang hilang dari pengurangan emisi karbon dioksida pada perkebunan kelapa sawit.* [Financial analysis and opportunity cost of carbon dioxide emission reduction in oil palm plantation]. *Jurnal Litbang Pertanian, 28*(4): 127–133 (in Indonesian).

Jaya, A. (2004). Ecological planning of tropical peatlands for carbon and water conservation. Ph. D. Thesis. University of Nottingham, UK (Unpublished).

Joosten, H., & Clarke, D. (2002). *Wise use of mires and peatlands – background and principles including a framework for decision-making*. Saarijärvi: International Mire Conservation Group & International Peat Society.

Kobayashi, N. (2008). *Sustainable peatland management: Socio-economic and environmental aspect*. Paper presented at international symposium "Sentinel earth: Detection of Environmental Change," at Hokkaido University, Sapporo, July 5–7.

MacKinnon, K., Hatta, M. G., Halim, H., & Mangalik, A. (2000). *Ekologi Kalimantan* [The Ecology of Kalimantan]. Jakarta: Prenhallindo.

Momose, K. (2002). Environments and people of Sumatran peat swamp forests II: Distribution of villages and interactions between people and forests. *Southeast Asian Studies, 40*(1), 87–108.

Noor, M. (2001). *Pertanian Lahan Gambut: Potensi dan Kendala* [Peat farming, the potential and limitations]. Yogyakarta: Kanisius (in Indonesian).

Noorginayuwati, A. Rafiq, R. Yanti, M. A., & Jumberi, A. (2006). *Penggalian Kearifan Lokal Petani untuk Pengembangan Lahan Gambut* [Exploring the farmers' local wisdom for peatland development]. Banjarbaru: Balai Penelitian Pertanian Lahan Rawa (in Indonesian).

Page, S., Morrison, E. R., Malins, C., Hooijer, A., Rieley, J. O., & Jauhiainen, J. (2011). *Review of peat surface greenhouse gas emissions from oil palm plantations in Southeast Asia*. Washington, DC: International Council on Clean Transportation.

Parish, F., Lim, S. S., Perumal, B., & Giesen, W. (Eds.). (2012). *RSPO Manual on Best Management Practices (BMPs) for management and rehabilitation of natural vegetation associated with oil palm cultivation on peat*. Kuala Lumpur: RSPO.

Putra, E. I., Hayasaka, H., Takahashi, H., & Usup, A. (2008). Recent peat fire activity in the mega rice project area, central Kalimantan, Indonesia. *Journal of Disaster Research, 3*(5), 1–8.

Rieley, J. O., & Page, S. E. (Eds.). (2005). Wise use of tropical peatlands: Focus on Southeast Asia. The Netherlands: Alterra. online at www.restorpeat.alterra.wur.nl.

Rina, Y., & Noorginayuwati. (2012). *Sosial dan ekonomi petani di lahan gambut* [Social and economic farmers on peatland]. In M. Noor, M. Alwi, D. Mukhlis, D. Nursyamsi, & M. Thamrin (Eds.), *Lahan Gambut: Pemanfaatan dan pengembangannya untuk pertanian* [Peatland: Its functions and development for farming] (pp. 217–247) (in Indonesian). Yogyakarta: Kanisius.

Ritung, S., Wahyunto, N. K., Sukarman, H., Suparto, C. T. (2011). *Peta Lahan Gambut Indonesia Skala* 1:250.000 (Map of Peatland in Indonesia, scale: 1:250,000, p. 11). Jakarta: Ministry of Agriculture, Republic of Indonesia (in Indonesian).

Saman, T., & Limin, S. (1999). Socio-economic values of wetlands for dayak community in Central Kalimantan. In T. Iwakuma et al. (Eds.) *Proceedings of the International Symposium on Tropical Peatlands, Bogor, Indonesia* (pp. 22–23). Sapporo: Graduate School of Environmental Earth Science, Hokkaido University.

Sastry, N. (2002). Forest Fires, air pollution, and mortality in Southeast Asia. *Demography, 39*(1), 1–23.

Schrier-Uijl, A. P., Silvius, M., Parish, F., Lim, K. H., Rosediana, S., & Anshari, G. (2013). *Environmental and social impacts of oil palm cultivation on tropical peat: A scientific review. Final report*. Kuala Lumpur: Roundtable on Sustainable Palm Oil.

Sheil, D., Casson, A., Meijpaard, E., Noordwijk, M. van., Gaskell, J., Sunderland-Groves, J., Wertz, K., & Kanninen, M. (2009). *The impacts and opportunities of oil palm in Southeast Asia: what do we know and what do we need to know?* (Occasional paper no. 51). Bogor: Center for International Forestry Research (CIFOR).

Starrett, D. A. (2000). Shadow pricing in economics. *Ecosystems, 3,* 16–20.

Tacconi, L. (2003). *Fires in Indonesia: Causes, costs and policy implications*. CIFOR (Occasional Paper No. 38).

Umar, S., Noor, M., & Noorginayuwati, A. (2014). Kearifan lokal untuk peningkatan dan keberlanjutan produksi pertanian di lahan gambut [in Indonesian; *The local wisdom for agricultural sustainability on peatland*]. Jakarta: IAARD.

Wetlands International. (2007). Assessment on peatlands, biodiversity and climate change. Kuala Lumpur: Global Environment Centre & Wageningen.

White, D., & Minang, P. (Eds.). (2011). *Estimating the opportunity costs of REDD +: A training manual*. Washington: The World Bank.

Whiteman, A., & Fraser, A. (1997). *The value of forestry in Indonesia*. Jakarta: Indonesia-UK Tropical Forest Management Programme.

WRI (World Resources Institute). (2005). *Ecosystems and human well-being: Synthesis*. Washington: Island Press.

Chapter 7
Agribusiness, Overdevelopment, and Palm Oil Industrial Restructuring in Malaysia

Kazuyuki Iwasa

Abstract Palm oil has become one of the most contested agro-commodities in developing countries. With characteristics of high productivity and general-purpose use, palm oil has been used worldwide for various items such as food, nonfood, and biofuel, and most frequently consumed in the oils and fats market. This global demand has been supported by a massive supply from Southeast Asia, mostly from Malaysia and Indonesia. Both countries have launched massive plantation development projects to become dominant world producers/exporters. However, in the shadow of this lucrative export boom, tropical ecosystems have been rapidly converted into vast monocultured landscape. This chapter examines the structure of the palm oil industry as a set of agro-industrial production linkages and then focuses on the recent transnationalization of agribusiness capital. It looks at and measures the latest transnationalization in a dual form within a short time, and explores the growth divergence between agribusiness and national industry/economy before building a substantial national economic base. Finally, it suggests that within agribusiness in Malaysia there is short-term economic orientation which influences how the national economy is built.

Keywords Agribusiness · Commodity chains · Global palm connections
Metabolism · Overdevelopment · Agro-business restructuring

This chapter derives in part from research results of the following JSPS joint researches; "Oil Palm Smallholders and Estate Corporations in Southeast Asia", and "An International Comparative Study on Agricultural Investment under the Restructuring of Agri-food Regime Investor Countries' Responsibility." I would like to thank Hayashida Hideki, Okamoto Masaaki, Hisano Shuji, and Darren Lingley.

K. Iwasa (✉)
Faculty of Humanities and Social Sciences, Kochi University, Kochi, Japan
e-mail: kiwasa@kochi-u.ac.jp

M. Lopez and J. Suryomenggolo (eds.), *Environmental Resources Use and Challenges in Contemporary Southeast Asia*, Asia in Transition 7,
https://doi.org/10.1007/978-981-10-8881-0_7

Introduction

Palm oil has become one of the most contested agro-commodities in developing countries. With characteristics of high productivity and general-purpose use, palm oil has been used worldwide for various items such as food, nonfood, and biofuel, and most frequently consumed in the oil and fat market. This global demand has been supported by a massive supply from Southeast Asia, mostly from Malaysia and Indonesia. Both countries have launched massive plantation development projects to become dominant world producers/exporters. However, in the shadow of this lucrative export boom, tropical ecosystems have been rapidly converted into a vast monocultured landscape. This transformation has been followed by environmental destruction, land disputes, widespread criticism, and severe controversy between business circles and socio-environmental NGOs (Wakker 2005; Colchester et al. 2006; Yusof 2008; Singh et al. 2009; Chao and Colchester 2012).

Reflecting an increasing global concern, palm oil has drawn growing attention from many researchers in terms of development versus the environment, and various studies have provided assessments of the diverse conditions of local communities and the implications of how palm oil production and trade can impact on global governance (McCarthy et al. 2012; Jiwan 2013; Pye 2013). While these are important issues to understand the changes, socioeconomic and environmental change should also be viewed in the context of the transnational behavior of agribusiness capital and related industrial restructuring (Iwasa 2011; Teoh 2013). This chapter focuses on the globalization of agribusiness capital and its social, economic, and ecological impacts in Malaysia, a pioneer in the large-scale development of palm oil.

Before presenting an analysis, this chapter offers two analytical viewpoints. First, it examines the structure of the palm oilindustry as a set of agro-industrial production linkages. Oil palm plantations do not stand alone, but include milling plants as fruits must be extracted within 24 hour after harvest. They are established on a large-scale allowing for the massive procurement of FFBs (fresh fruit bunches) to secure plant profitability. Therefore, the smooth transportation of FFBs from plantations to mills is an imperative, and after producing crude palm oil, the oil is subsequently transported from mills to refineries to produce various processed oils for the world market. This chapter views a set of production/labor linkages from a commodity chains perspective (Gereffi and Korzeniewicz 1994; Fold and Prichard 2005; Bair 2009), and explores agro-industrial interrelationships and restructuring between upstream and downstream sectors.

The second focus is on the recent transnationalization of agribusiness capital. While the growth rate of Malaysian palm oil production has slowed down since the 1990s, Malaysian agribusinesses have increasingly moved across the border and reorganized existing commodity chains. This transnationalization consists of the following patterns: capital invests abroad to access overseas markets and resources, and the importation of foreign labor to combine with domestic resources. This trend is expected to transform the metabolism between nature and human beings, and will have significant economic, sociopolitical, and ecological effects on both host and home countries[1]. This chapter measures the latest transnationalization in a dual form within a short time and explores the growth divergence between agribusiness and national industry/economy before building a substantial national economic base. It suggests that, within agribusiness in Malaysia, there is short-term economic orientation which influences how the national economy is built.

This chapter comprises of three parts. First, it presents the development trajectory of the Malaysian palm oil industry and its recent slowdown, and describes the recent global strategies of the largest palm oil-related agribusinesses in Malaysia. After that, it will describe the industrial restructuring that has taken place within the Malaysian palm oil industry and its economic, sociopolitical, and ecological impacts in terms of the following viewpoints: the extension of overdevelopment and sociopolitical disruption; international competition and commodity chains restructuring; and the importing of foreign labor force and agricultural vulnerability. Last, based on these descriptions, it will draw some conclusions and prospects for the industry.

Development Trajectory of the Malaysian Palm Oil Industry

Extensive Development of Large-Scale Plantations

At an early stage in Southeast Asia, Malaysia was the most successful country in introducing large-scale oil palm plantation development. From the 1960s, the price of natural rubber—once the country's prime export commodity—slumped and oil palm came to be regarded as an alternative crop in terms of similar cultivation conditions and consistent with policy recommendations made by the World Bank and the Ford Foundation (Khera 1976; Iwasa 2005). Planting was pushed forward in line with agricultural diversification. As a result of this crop selection and concentration, land area given to oil palm increased from 55,000 ha in 1960 to 1 million ha in 1980, 2 million ha in 1990 to over 4 million ha in 2005, and 5.6 million ha in 2015. This makes oil palm the largest crop in the country, accounting for 60% of total arable area. Furthermore, the area in which oil palm is planted has also spread across the country through extensive development of newly created plantations. In conjunction with its expansion, palm oil production dramatically

---

[1]Metabolism (*Stoffwechsel*) means the circulation process of matter between nature and human beings through labor. Human beings act upon external nature and change it to appropriate the natural materials in a form adapted to their needs, and they simultaneously change their own nature. However, modern capitalist agriculture "disturbs the metabolic interaction between man and the earth," brings ecological rift, and destroys both nature and human health (Marx 1867/1977, pp. 636–639; Foster 2000).

increased over 200 times from 92,000 tons to nearly 20 million tons during the same period (Table 7.1).

The rush to develop this market has made Malaysia the world's leading producer/exporter. Table 7.2 shows data for the transition of top five palm oil producing/exporting countries. Malaysia was in a mere fourth position in 1961, but from the 1970s to 1990s it surpassed West African countries, oil palm's place of origin, to become the top producer/exporter. But in the 2000s, as will be discussed, Indonesia surpassed Malaysia as the prime supplier.

In the process of development, many palm oil planters have entered into the field of palm oil cultivation, following three types of operation (MPOB 2014). The first one is private companies, which occupy 62% share of the total area in 2013. Most companies originated in the former British Agency Houses, but, after the late 1970s and the early 1980s, became localized through stock acquisition by PNB *(Permodalan Nasional Berhad)*, a government holding company, or by politically connected Malaysian ethnic Chinese (Teoh 2013; Tan 2008). The second type of operation is smallholders organized by Federal/State government agencies, accounting for a quarter of the total area. This type represents farmers under the control of government organizations, such as FELDA (Federal Land Development Authority) for the purpose of poverty eradication of the landless poor. The third type is independent smallholders, which occupy 14% of the total area. Among these three types, private companies including Sime Darby, KL Kepong or the IOI Corporation and government-linked FELDA have topped the list and served as important driving forces in palm oil production of the country.

Agro-Industrialization and Transformation into NACs

As mentioned above, the palm oil industry consists not only of the plantation sector itself but of subsectors including the following: (1) the milling sector which extracts crude palm oil from FFBs harvested in plantations; (2) the refining sector which refines, breaches, and deodorizes crude oil to manufacture processed palm oil and finished products; (3) the oleochemical sector which produces fatty acids and alcohols for using soap, detergent, and cosmetics, etc.; and (4) the biodiesel sector which manufactures fuel for transportation (Fig. 7.1).

In fact, along with the increase of upstream production volume, downstream sectors have also successively developed in line with the palm oil commodity chain. As seen clearly in Table 7.3, the milling sector was established in the 1960s, and there is a significant increase in the number of plants in the refinery sector (from the 1970s), the oleochemical sector (from the 1980s), and the biodiesel sector (from the 2000s). In the process of industrial growth, large-scale Malaysian plantation companies and foreign-affiliated enterprises or joint venture companies have had a strong presence in downstream sectors. For example, the refinery sector consists of three types: (1) vertical integrated Malaysian companies (e.g., Sime Darby, IOI Corporation, FELDA); (2) nonintegrated Malaysian companies with small or no

Table 7.1 Oil palm planted area and palm oil production in Malaysia

	Total	Planted area (000 ha)			Percentage of planted area (%)				Production of crude palm oil (000t)	5-year growth rate (%)	
					Total						
		Peninsular Malaysia	Sabah	Sarawak		Peninsular Malaysia	Sabah	Sarawak		Planted area	Production
1960	55	–	–	–	–	–	–	–	92	–	–
1965	97	–	–	–	–	–	–	–	150	76.4	63.0
1970	291	261	29	1	100.0	89.7	9.9	0.4	431	200.0	187.3
1975	642	569	59	14	100.0	88.6	9.2	2.2	1258	120.5	191.8
1980	1023	907	94	23	100.0	88.6	9.2	2.2	2573	59.4	104.6
1985	1482	1292	162	29	100.0	87.2	10.9	1.9	4134	44.9	60.7
1990	2029	1698	276	55	100.0	83.7	13.6	2.7	6095	36.9	47.4
1995	2540	1903	518	119	100.0	74.9	20.4	4.7	7811	25.2	28.2
2000	3377	2046	1001	330	100.0	60.6	29.6	9.8	10,842	32.9	38.8
2005	4051	2299	1209	543	100.0	56.7	29.9	13.4	14,962	20.0	38.0
2010	4854	2525	1410	919	100.0	52.0	29.0	18.9	16,994	19.8	13.6
2015	5643	2659	1544	1439	100.0	47.1	27.4	25.5	19,962	16.3	17.5

Note: Planting started from 1959 in Sabah, and from 1969 in Sarawak

Source: MPOB (2014). *Malaysian oil palm statistics 2013*; Dept. of Statistics Malaysia, *Oil palm, coconut and tea statistics 1975, 1970 and 1969;* MPOB Economics & Industry Division Website (http://bepi.mpob.gov.my/. Accessed May 22, 2016)

Table 7.2 Top five countries of palm oil production/exports

(▪, figures in parentheses are million tonnes)

	1961		1970		1980		1990		2000		2013	
Production	World total	100.0 (1.5)	World total	100.0 (1.9)	World total	100.0 (5.1)	World total	100.0 (11.4)	World total	100.0 (22.2)	World total	100.0 (55.8)
	Nigeria	45.2	Nigeria	25.2	**Malaysia**	**50.6**	**Malaysia**	**53.2**	**Malaysia**	**48.8**	Indonesia	50.9
	Democratic Republic of the Congo	15.1	**Malaysia**	**22.3**	Indonesia	14.2	Indonesia	21.1	Indonesia	31.5	**Malaysia**	**34.4**
	Indonesia	9.9	Democratic Republic of the Congo	12.0	Nigeria	12.8	Nigeria	6.4	Nigeria	4.0	Thailand	3.5
	Malaysia	**6.4**	Indonesia	11.2	Côte d'Ivoire	3.7	Colombia	2.2	Thailand	2.6	Nigeria	1.7
	Angola, China	2.7	China	5.8	Democratic Republic of the Congo	3.3	Thailand	2.0	Colombia	2.4	Colombia	1.7
Export	World total	100.0 (0.6)	World total	100.0 (0.9)	World total	100.0 (3.6)	World total	100.0 (8.1)	World total	100.0 (14.2)	World total	100.0 (41.7)
	Nigeria	26.6	**Malaysia**	**44.4**	**Malaysia**	**59.1**	**Malaysia**	**70.1**	**Malaysia**	**57.5**	Indonesia	49.4
	Democratic Republic of the Congo	24.5	Indonesia	17.6	Singapore	18.1	Indonesia	13.6	Indonesia	29.0	**Malaysia**	**36.6**
	Indonesia	18.6	Singapore	14.7	Indonesia	14.1	Singapore	7.7	Papua New Guinea	2.4	Netherlands	3.8
	Malaysia	**15.1**	Democratic Republic of the Congo	13.1	Côte d'Ivoire	2.7	Côte d'Ivoire	1.9	Netherlands	2.3	Papua New Guinea	1.4
	Singapore	4.7	Netherlands	2.1	Netherlands	2.1	Netherlands	1.5	Singapore	1.2	Thailand	1.3

Note: Data is based on volume.
Source ▪ FAO, *FAOSTAT* ▪ Accessed May 20, 2016 ▪.

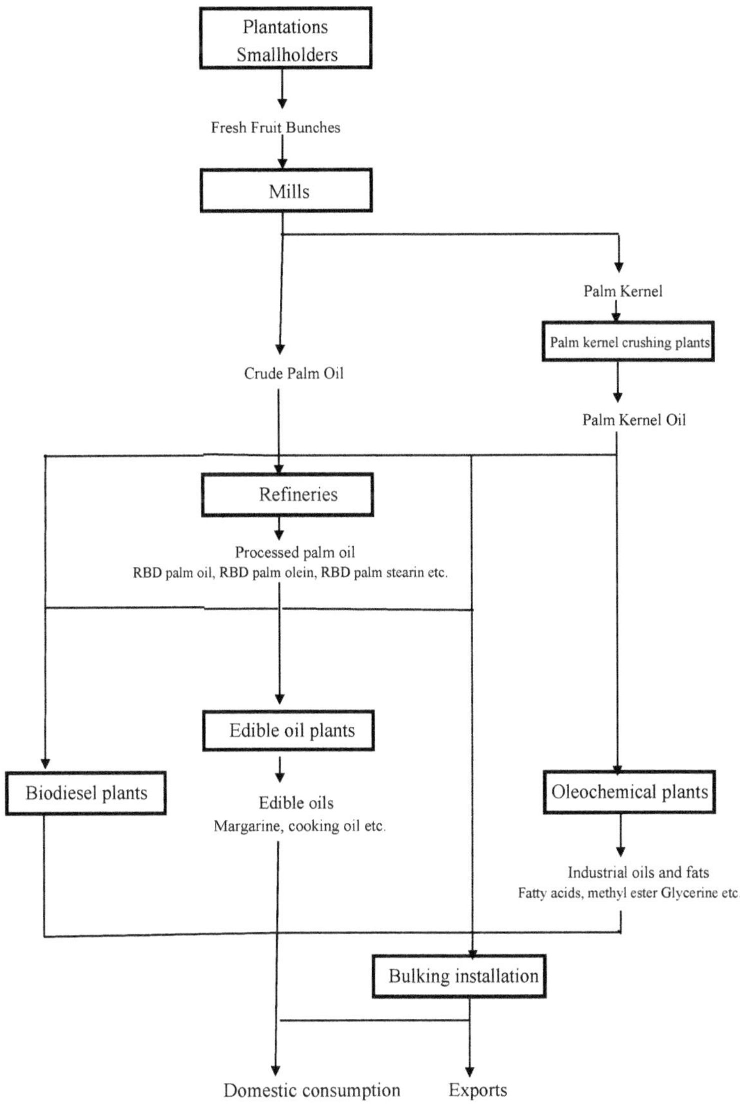

Fig. 7.1 Conceptual diagram of palm oil commodity chains

plantations (e.g., Lam Soon, Southern); and (3) independent foreign-affiliated companies (e.g., Wilmar, Cargill, ISF). Furthermore, in the oleochemical sector, joint ventures between vertically integrated plantation companies and multinational corporations have become mainstream (Iwasa 2005; Interview with PORAM, August 13, 2012).

This dynamic agro-industrialization has had various effects on both national and international spheres. First, agroindustry has brought about export expansion both

Table 7.3 Development of palm oil-related industrialization in Malaysia

	Mills		Refineries		Oleochemical plants		Biodiesel plants	
	No. of plants	Capacity (000t/FFB/year)	No. of plants	Capacity (000t/CPO/year)	No. of plants	Capacity (000t/year)	No. of plants	Capacity (000t/year)
1975	82	n.a.	8	n.a.	–	–	–	–
1980	149	19,594	45	2880	–	–	–	–
1985	229	35,122	38	5350	–	–	–	–
1990	261	42,874	37	10,454	–	–	–	–
1995	281	50,798	41	10,147	13	824	–	–
2000	350	65,949	46	14,599	16	1800	–	–
2005	397	86,244	51	18,506	17	2467	–	–
2010	421	97,386	51	22,886	18	2599	18	2272
2014	443	106,708	57	27,051	18	2665	18	2860

Note: Described operating plants only
Data of oleochemical plants is available after 1995. Biodiesel plants started operation from 2006

Source: MPOB (2015). *Status perkumbangan industri 2014* (http://econ.mpob.gov.my/economy/mdustry2/profile2014/profile2014.html); MPOB, *Malaysian oil palm statistics*, various issues; PORLA, *Palm oil statistics*, various issues; Dept. of Statistics Malaysia *Oil palm, coconut and tea statistics, 1975*

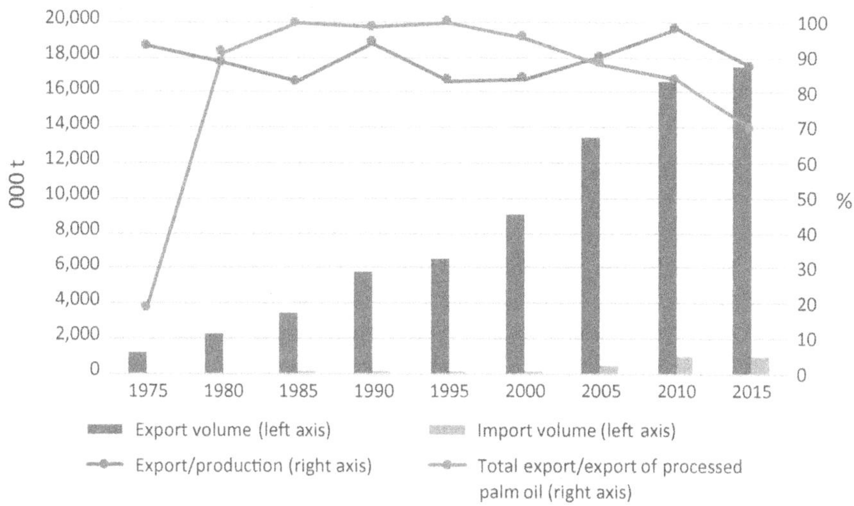

Fig. 7.2 Export/Import trends of Malaysia palm oil *Source* MPOB (2014). *Malaysian oil palm statistics 2013*; MPOB Economics & Industry Division Website (http://bepi.mpob.gov.my/. Accessed May 22, 2016)

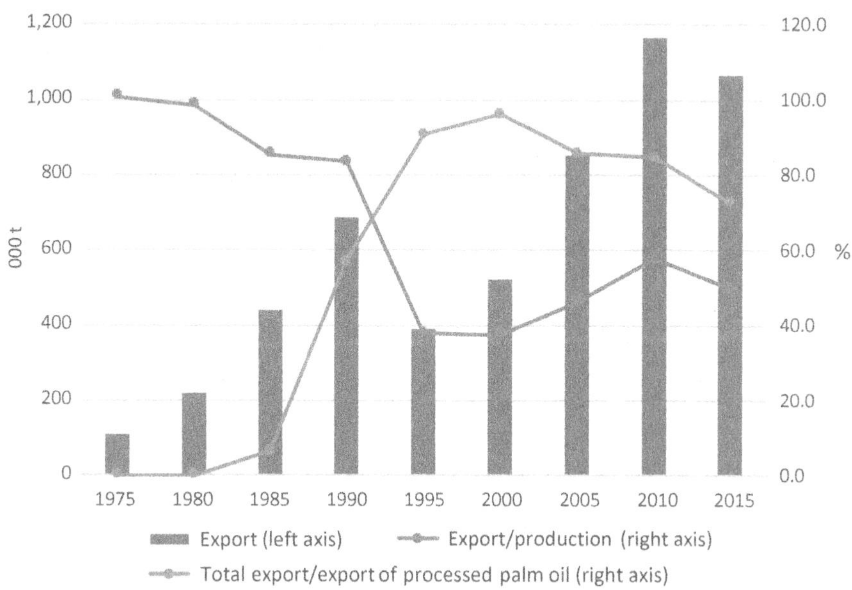

Fig. 7.3 Export trends of Malaysian palm kernel oil *Source* MPOB (2014). *Malaysian oil palm statistics 2013*; MPOB Economics & Industry Division Website (http://bepi.mpob.gov.my/. Accessed May 22, 2016)

in terms of quantity and quality. For four decades, between the years of 1975–2015, palm oil export volume increased 15 fold from 1.2 million tons to 17 million tons, accounting for about 90% of the total production. Processed palm oil export rapidly outnumbered crude palm oil export, and the percentage of processed palm oil in total exports increased from 18% to 91% during the late 1970s (Fig. 7.2).

In the process of increasing the share of processed palm oil, Malaysia has exported a wide variety of palm oil products, in response to the demand from the global market. For example, export to China is mainly liquid RBD palm olein for cooking oil, while the main product to South Asia is semisolid RBD palm oil as a raw material for local ghee. Palm kernel oil export also increased 10-fold and the percentage of processed palm kernel oil in total exports grew significantly during 1975–2015, but from the 1990s, the ratio of export to production decreased below a half, which indicates the transition from direct oil export to domestic processing for oleochemical plants (Fig. 7.3). Thus, what is clear is that the development of palm oil industrial commodity chains has diversified and upgraded for market export.

Second, the palm oil industry has undoubtedly contributed to national economic development. It is the fourth largest industry among GNI contribution, and identified as one of the three major sectors with the oil/gas sector and the financial services sector under the Economic Transformation Program (PEMANDU 2010). In the manufacturing sector, the industry accounts for 12% of total manufacturing output in the country, and crude and refined palm oil subsectors are the second and the third largest industrial subcategories respectively in 2011 (Department of Statistics Malaysia 2012). In terms of external trade, palm oil consists of less than 10% of the total export value, and 32% of the trade surplus in 2014 (Department of Statistics Malaysia 2015). In contrast to the electronics industry, which is the country's top export but with a low trade surplus due to the high parts/components import, palm oil serves as an important source of foreign exchange based on the high level of local contents. Considering its successful export-oriented development trajectory, Malaysia can be defined as one of the New Agricultural Countries (NACs) based on its emerging palm oil export.[2]

Slowdown in NACs Model Development

In recent years, however, new problems have emerged against a backdrop of undermining domestic production base as well as the emergence of a neighboring competitor. There are, at least, two main concerns here.

First, Malaysia is facing a land shortage for plantations. Area expansion has been performed by extensive development of newly created plantations in conjunction with the intensive development of rubber trees replacement. At an early stage,

[2]In the literature of political economy of food and agriculture, some developing countries based on export-oriented agriculture are referred to as NACs (New Agricultural Countries) to differentiate them from NICs (Newly Industrializing Countries) (McMichael 1996).

Peninsular Malaysia was a production center, but along with limitations to land cultivation, main production areas have shifted to Sabah and Sarawak in East Malaysia. As a result, the area share of both states has dramatically increased from 16% to 53% during the period 1990–2015. The growth rate of planted areas is decreasing except in Sarawak, as illustrated in Table 7.1. Excluding the amount of agricultural land, production yields are also sluggish especially in Peninsular Malaysia, which is dependent on old trees necessary to be replanted.

Second, Malaysia has also a critical problem relating to its rural labor force. Due to ongoing industrialization and growth of the service economy, there has been an outflow of the rural population to urban areas. According to the population census, the ratio of urban population increased from 34% in 1980 to 51% in 1991, and 71% in 2010 (Department of Statistics Malaysia 2011). In particular, field operations in the oil palm industry are highly labor-intensive. The rural exodus has meant that there has been a growing labor shortage on many plantations.

As a result of this shortage of both land and labor, the growth rate of palm oil production has gradually declined from 38% in 2000–2005 to 14% in 2005–2010, and 18% in 2010–2015 (Table 7.1). Coupled with sluggish growth and the development of large-scale plantations in Indonesia, Malaysia ceded top palm oil production/export to Indonesia from 2006 and 2009 respectively, which led to increased international competition, as illustrated in Table 7.2. Although Malaysia remains the main exporter, Malaysian palm oil imports have shown an increase since the 2000s (Fig. 7.2). What is implied here is that the Malaysian palm oil industrial base is weakening, and the NAC development model is showing signs of slowing down.

Global Strategies of Palm Oil-Related Agribusinesses

Oil Palm Development and Appearance of Large-scale Agribusinesses

This section focuses on the operations and strategies of agribusiness capital, a driving force of development. First of all, it offers an overview of leading oil palm plantation companies worldwide.

Table 7.4 shows the composition of the leading palm oil plantation companies in the world, which indicates the following characteristics. First, Southeast Asian companies occupy the top rankings. The number of home countries is Malaysia (11), Indonesia (9), and Singapore (3) respectively. The largest company is Malaysia-based Sime Darby, followed by Golden Agri-Resources (GAR, based in Singapore, but an affiliated business of Indonesia-based Sinar Mas), Felda Global Ventures Holdings (Malaysia), PT Astra Agro Lestari (Indonesia), Wilmar International (Singapore, but co-founded by Singaporean Kuok Khoon Hong and Indonesian Martua Sitorus), and PT Salim Ivomas Pratama (Indonesia). It should be

Table 7.4 Leading palm oil plantation companies in the world

Industry players	Country of incorporation	Location of plantations	Planted area (000 ha)	FFB production (000t)	Annual FFB Yield (t/mature
Sime Darby Berhad	Malaysia	Malaysia, Indonesia and Liberia	521.9	10,048.0	21.5
Golden Agri-Resources Ltd	Singapore	Indonesia and Papua New Guinea	455.7	8508.7	21.8
Felda Global Ventures Holdings Berhad	Malaysia	Malaysia	323.6	5197.3	19.9
PT Astra Agro Lestari Tbk	Indonesia	Indonesia	266.7	4798.5	n.a.
Wilmar International Limited	Singapore	Indonesia, Malaysia, Uganda and West Africa	247.1	4073.0	19.8
PT Salim Ivomas Pratama	Indonesia	Indonesia	216.8	2797.0	17.7
Kuala Lumpur Kepong Berhad	Malaysia	Malaysia and Indonesia	187.1	3289.0	22.2
Asian Agri Group[a]	Indonesia	Indonesia	160.0	n.a.	n.a.
IOI Corporation Berhad	Malaysia	Malaysia	158.2	3295.5	23.7
PT Sinar Mas Agro Resources and Technology Tbk	Indonesia	Indonesia	139.0	2742.2	21.9
PT Perkebunan Nusantara IV (PERSERO)[a]	Indonesia	Indonesia	135.3	2191.4	n.a.
First Resources Limited	Singapore	Indonesia	132.3	1898.6	22.2
PT Bakrie Sumatera Plantations Tbk[a]	Indonesia	Indonesia	115.8	n.a.	n.a.
Kulim (Malaysia) Bhd[a]	Malaysia	Malaysia, Papua New Guinea and the Solomon Islands	112.2	2375.4[b]	21.7
PT Perkebunan Nusantara III[c]	Indonesia	Indonesia	104.9	1629.9	24.2
PT Sampoerna Agro[a]	Indonesia	Indonesia	102.8	1376.4	15.6–19.4
Tradewinds Plantations Berhad[a]	Malaysia	Malaysia	91.1	1191.7	16.9

(continued)

Table 7.4 (continued)

Industry players	Country of incorporation	Location of plantations	Planted area (000 ha)	FFB production (000t)	Annual FFB Yield (t/mature
Genting Plantations Berhad[a]	Malaysia	Malaysia and Indonesia	89.1	1372.0[b]	21.2
Socfin Group[a]	Luxembourg	Malaysia, Indonesia, Côte d'Ivoire, Nigeria and Cameroon.	86.5	n.a.	n.a.
PT PP London Sumatra Indonesia Tbk	Indonesia	Indonesia	80.7	1291.3	18.4
Boustead Holdings Berhad	Malaysia	Malaysia and Indonesia	74.2	1121.6	17.1
United Plantations Berhad	Malaysia	Malaysia and Indonesia	45.7	n.a.	20.7
IJM Plantations Berhad	Malaysia	Malaysia and Indonesia	38.8	575.2	23.7
Hap Seng Plantations Holdings Berhad	Malaysia	Malaysia	37.1	739.0	23.8

Note: Data of Felda Global Ventures Holdings included only direct management

Data acquisition year regarding companies with alphabetic characters is as follows; [a]2010, [b]2011, [c]2009

Source: FGVH (2012). *Prospectus* (Extracted from Frost & Sullivan's independent market research reports)

noted that each of these companies manages large-scale planted areas which are over 200,000 ha respectively.

Second, Indonesia is increasingly becoming an important oil palm development center. Besides Indonesian domestic capital, Malaysian and Singaporean foreign capitals have also seen high levels of investment in Indonesia for developing large-scale plantations. This has been a driving force of Indonesia allowing it to become the largest palm oil producer/exporter in the world.

Third, the location of oil palm plantations has extended outside Southeast Asia. Especially, some Malaysian and Singaporean agribusinesses have set up plantations in Papua New Guinea and the Solomon Islands in Oceania as well as well as in West African countries. What we see here is a definite trend toward the globalization not only of palm oil trade but also of oil palm development investment.

Vertical Integration and Transnationalization of Malaysian Agribusinesses

Through a detailed analysis of the top five oil palm plantation companies operating in Malaysia such as Sime Darby, FELDA, Kuala Lumpur Kepong, IOI Corporation, and Wilmar International, this chapter now turns to some of the latest features of the structure and dynamics of agribusiness (Table 7.5).

First, every agribusiness has not only been a large-scale plantation company but has also diversified its oil and fat business through vertical integration. Based on the massive procurement of FFBs in plantations, each company has moved to the high value-added downstream sectors such as mills, refinery, oleochemical, and biodiesel plants. Remarkably, they have extended their businesses by using multiple strategies such as M&A and set up of joint ventures with transnational corporations.

For example, in the refinery sector, FELDA have established joint ventures with Mitsui Co./ADEKA (Japan) and IFFCO (UAE), and IOI took over Pan Century, (India) and Loders Croklaan, Unilever's subsidiary of leading chocolate fats manufacturing, for the purpose of expanding their export market and moving to the higher value-added business. Similarly, in the oleochemical sector, FELDA and Sime Darby set up joint ventures with P&G (U.S.) and Cognis (Germany) respectively, and IOI bought up Palmco and Pan Century, aiming at acquiring technological know-how. KL Kepong created the largest plant worldwide with Japanese Mitsui, ADEKA, and Miyoshi. By using these strategies, every company now holds the entire sector of the palm oil commodity chains.

Second, all agribusinesses have not only built alliances with transnational corporations but also set up themselves as transnational corporations. Two distinctive patterns are depicted. First is overseas downstream business deployment such as China, India, and EU. For example, Wilmar has multiple manufacturing subsidiaries and associates in as many as nearly 200 plants worldwide for a leading brand-name cooking oil in China such as Arawana (blended oils), Koufu (blended

Table 7.5 Vertical integration and transnationalization of major palm oil-related agribusinesses in Malaysia

Name of companies	Corporate overview	Operation in Malaysia		Overseas operation	
		Plantations and mills sector	Refinery, oleochemical, specialty fats, and biodiesel sector	Plantations	Refinery, oleochemical, specialty fats, and biodiesel sector
Sime Darby	Established in 2007 by merging Golden Hope, Kumpulan Guthrie, Sime Darby under PNB. Largest public listed company, and world's largest agriculture-based and plantation TNC ranked by foreign assets	• Plantatin 130. Total about 350,000 ha (Oil palm 305,000ha) • Mills 34. FFB processing over 10 million t/year. Production of crude palm oil 2.45 million t/year (including overseas operations)	• 4 refinery companies • Set up of JV with Cognis, German company • Sime Darby Biodiesel (operating 2 plants in Selangor)	• Indonesia: Minamas Plantation (282,045ha, 25 strategic operating units). Sime took over 4 plantation companies of Salim group in the Economic Crisis of 1997 • Liberia: Sime Darby Plantation (Liberia) Inc. (220,000ha). 63-year concession agreement with the Government of Liberia • Papua New Guinea & the Solomon Island:New Britain Palm Oil Limited (NBPOL) (136,268ha). oil palm-over 80,000ha, sugar-over 5,700ha, grazing pasture-9,145ha, 12 oil mills • Cameroon undertaking land deals (300.000ha)	• Indonesia PT Nusantara • Thailand Morakot Industries, • Vietnam Golden Hope Nha Be Edible Oils • Netherlands Sime Darby Unimills (acquisition from a Unilever's subsidiary in 2002) • Papua New Guinea and the Solomon Island New Britain Palm Oils • South Africa Sime Darby Hudson & Knight • Establishment of Golden Hope Bodiesel in Netherlands
FELDA	Founded in 1956. Government agency of land development and settlement for the purpose of rural landless poor. Currently FELDA ceased settlement for the poor and only manages own plantation. On October 3, 2003, FELDA established Felda Holdings Bhd., a public company with authorized capital of 5 billion ringgit. Felda Global Ventures was formed in 2007 for overseas operation and launched initial public listing in 2012	• Plantation 420,000ha (Oil Palm 305,000ha). FELDA settlers' area 370,000ha • Mills 72. Kernel crushing plants 4. Production of palm oil 3 million t/year	• 4 refineries, including a joint venture with Mistui Co • FPG Oleochemicals, a JV with P&G • Set up of FGV Green Energy, a JV with U.S. Benefuel	• Acquisition of PT Citra Niaga in West kalimantan (14,000ha) • Attempt deal for Indonesia's Rajawali Group in 2015 (The plan was cancelled and alleged improprieties emerged) • Plan to invest in Mindanao, the Philippines	• Refinery and bulking installations in Indonesia (refinery), China (Voray Holdings), and Pakistan (MEO). FELDA Iffco (JV with a UAE company) deploys refinery plants in Indonesia, Turkey etc • Marketing offices in Indonesia, UAE, France, Spain • Take over Twin River, U.S. based biodiesel company in 2007

(continued)

Table 7.5 (continued)

Name of companies	Corporate overview	Operation in Malaysia		Overseas operation	
		Plantations and mills sector	Refinery, oleochemical, specialty fats, and biodiesel sector	Plantations	Refinery, oleochemical, specialty fats, and biodiesel sector
IOI Corporation	Founded in 1969, and real estate company from 1982. Chinese capital. Since 1985 IOI started oil palm plantation activities by taking over plantation company. 44th largest agriculture-based and plantation TNC in the world by foreign assets. Lee Shin Cheng, IOI's executive chairman, is the Malaysia's 5th richest in 2016, holding 4.4 billion dollars	• Plantation 160,000ha (Peninsular Malaysia 43,000ha, East Malaysia 116,000ha). Acquisition of Unilever's plantations (Unipamol) in 2003. Tissue culture laboratory for breeding high-yield clones • Mills 14. FFB processing 4.75 million t/year (including operations in Indonesia)	• Acquisition of Palmco in 2001. Merger of Pan Century (Indian Company) • Buyout of Unilever's Loders Croklaan (specialty fats) in 2002 • Acquisition of Palmco in 2001. Fatty Chemical Malaysia, a JV with Kao • IOI Biodiesel	Investing in Indonesia since 2007. 31.7% stock control of Bumitama Agri (120,000ha)	• Acquisition of Loders Croklaan (a Unilever's subsidiariy) in 2002. Currently operating the largest palm refinery business in Europe • IOI Oleo GmbH in Germany, operating oleochemical plants • Non trans-fatty acid manufacturing business in U.S. and Canada • Specialty fats business in China • Plant construction with 200,000t capacity in the form of JV with BioX group in Rotterdam
KL Kepong	Founded in 1906, as London-based Kuala Lumpur Rubber Company Limited. After the crisis in 1969, Kuala Lumpur Kepong Berhad was set up in Malaysia and Lee Loy Seng was appointed the Founder Chairman in 1973. Today, 7th largest agriculture-based and plantation TNC in the world ranked by foreign assets. Lee Oi Hian & Lee Hau Hian, KLK's CEO, are the Malaysia's 15th richest in 2016, holding 915 million dollars	• Plantation 110,000ha (Peninsular Malaysia 69,000ha, Sabah 40,000ha) • Mills 14	• 2 Refinery plants • KLK Oleo (JV with Palm Oleo, ADEKA, Miyoshi), world's largest fatty acid producer •KLK Bioenergy	• Investing in Indonesia since 1994, controlling 137,486ha. JV with PTPN II to form PT LNK in 2009 • Liberia (since 2013): 21,018ha, JV with Liberian Palm Developments Limited	• Mills: Indonesia 9 • Refinery: Indonesia 3 • Oleochemical: KLK Oleo (Shanghai) in China. KLK Emmerich in Germany, acquisition of the German company Rheinsee 311.V V GmbH and merger with Uniqema in 2010. KLK Tensachem SA in Belgium. KOLB in Switzerland

(continued)

Table 7.5 (continued)

Name of companies	Corporate overview	Operation in Malaysia		Overseas operation	
		Plantations and mills sector	Refinery, oleochemical, specialty fats, and biodiesel sector	Plantations	Refinery, oleochemical, specialty fats, and biodiesel sector
Wilmar International	Established in 1991 as Singapore based palm oil trading company by Kuok Khoon Hong and Martua Sitorus, Indonesian tycoon. It made an acquisition of Kuok Group's PPB in 2007 with 60,000ha plantations and 7 mills. It is 21st largest agriculture-based plantation TNC in the world by foreign assets. ADM, a US based grain giant, made a capital participant. Kuok Khoon Hong, Wilmar's CEO and nephew of Malaysian billionaire Robert Kuok (Wilmar's largest shareholder), is the Singapore's 9th Richest in 2016, holding 2.5 billion dollars	• Plantation 60,000ha (Sabah and Sarawak) • Mills 7	• 6 refinery plants • 1 oleochemical plant • 1 specialty fats plant • 2 biodiesel plants (PGEO Bioproducts)	• Indonesia 180,000ha in Sumatra, West Kalimantan and Central Kalimantan. Plasma scheme 38,000ha • JVs in Uganda (6,000ha and outgrowers 300ha) and West Africa(39,000ha and outgrowers 140,000ha) • Côte d'Ivoire (47,000ha), Ghana (6,000ha)	• Mills and refineries in China (51), Europe (4), Vietnam (2) etc • Oleochemical business in China (7), Indonesia (1) • Specialty fats business in Indonesia (2), Europe, Vietnam etc • Oils and fats business in Russia, India, Ukraine, Bangradesh • Manufacturing original-brand cooking oils in China, India, Indonesia

Note: Described only palm oil-related business

Source: Each company's *Annual Report* and corporate website; UNCTAD (2009). *World Investment Report 2009*. United Nations; Forbes "Malaysia's 50 richest" (https://www.forbes.com/malaysia-billionaires/list)

oils), and Neptune (palm olein), and for the original-brand cooking oil in India (e.g., Raag and Alpha) and Indonesia (e.g., Sania and Fortune) respectively. KL Kepong also undertakes oleochemical business in EU through its acquisition of a German Rheinsee 311.VV plant to become a global oleochemical manufacturer. IOI holds the EU's largest refinery in the Netherlands, and has set up non-trans-fatty acid manufacturing businesses in North America. What is clear here is that creating locational advantages in the main export market is a common strategy among these companies.

Another pattern is cross-border plantation development. Since the 1990s, agribusinesses have extended their development space from East Malaysia to neighboring Indonesia in search of vast tracts of land for oil palm plantations. As a result, in the case of KL Kepong and Wilmar, planted areas in Indonesia have become larger than that of Malaysia. Through these transnational strategies, agribusinesses have sought to break through the restrictions of land and labor problems and to extend the palm oil commodity chains across borders.

Furthermore, after the 2000s, against the rise of Indonesian plantation companies and increased international competition, each agribusiness has pushed forward global business strategies. The first is a restructuring of the plantation sector. In 2007, three major corporations of Sime Darby, Golden Hope, and Kumpulan Guthrie merged into a single entity as Sime Darby Berhad, the largest agriculture-based TNC ranked by foreign assets in the world (UNCTAD 2009). In the same year, Wilmar made an acquisition of PPB group, Kuok group's oil-and-fat affiliated company. On the other hand, plantation development has globalized. Sime Darby Berhad has, for example, advanced into Liberia, Papua New Guinea, and the Solomon Islands. In the case of Liberia, Sime Darby Berhad has launched over 200,000 ha of oil palm plantation development on a long contract of 63 years (Sime Darby Plantation Sdn Bhd n.d.). Coping with a forest moratorium in Indonesia, the conglomerate has focused on West Africa to explore new sources for the EU biodiesel market (Levitt 2011). Wilmar has also expanded into Africa like Uganda and West African countries for developing new oil palm plantations. In short, due to the global demand for palm oil as well as international scrutiny/national regulation of forest clearance and limitation of expanding oil palm plantations nationally, development space has recently been expanded from Peninsular Malaysia across East Malaysia/Indonesia to outside Southeast Asia. Today, new circuits of production, manufacturing, distribution, and consumption globally, that is, palm oil-related global commodity chains or "global palm connection," are undergoing rapid configuration.

Second is the public listing of FELDA group. FELDA is a government agency for eradicating poverty through land development and resettlement of the rural landless poor. As the scale of development has become larger, however, it has expanded its commercialized downstream sectors, and become the world's largest palm oil manufacturer, contrary to withdrawing from the settlement project for the poor (Iwasa 2005). FELDA established Felda Global Ventures Holdings in 2007, and further launched initial public listing, which gained U.S. $3,200 million in June 2012. This was the world's second-biggest initial public offering in that year after

Facebook (Yantoultra and Niluksi 2012). Furthermore, it is reported that Felda Global sets the goal of increasing its revenue to RM100 billion ($31 billion) by 2020, an eightfold increase compared with 2013, to be one of the top 10 agribusiness conglomerates in the world through restructuring of the FELDA group and affiliate companies' stock listings (Tan 2014: Felda Global Ventures Holdings Berhad 2016). This process means that FELDA has fully transformed itself from the national development organization into transnational agribusiness capital with high-profit palm oil business.

Third, some agribusinesses have formed strategic alliances with the world's largest transnational grain traders. For example, Wilmar was made a capital participant by the U.S. agribusiness giant ADM from the 1990s to set up JV in China, designed to build a collaborative relationship regarding soybean oil and palm oil. In October 2012, they have signed regulatory approvals for their partnerships in global fertilizers and European vegetable oils (ADM 2012). Felda Global also established a partnership with Louis Dreyfus and Bunge, global agribusinesses like ADM and Cargill. In the case with Bunge, Felda Global set up a joint venture for the oilseed business like soybean and canola in Canada, and is expected to gain tolling fees from Bunge based on the latter's usage of its downstream facilities. Assuming the market risk of the soybean and canola purchasing and hedging, however, Felda Global terminated its joint venture in November 2013 (Borneo Post 2013). IOI also announced to sell out 70% controlling stake in Loders Croklaan to Bunge for RM3.79 billion in 2017 for the purpose of reducing debt and using Bunge's global network (Adnan 2017). What these cases indicate is that both Malaysian agribusiness and grain traders are trying to engage in dealing with multiple vegetable oils including palm oil, soybean, and canola, and through a strategic partnership they will extend their oligopolistic status in the world oils and fats market.

Under the recent strategies of vertical integration and transnationalization, agribusinesses have reaped bountiful fruits from palm oil business. As shown in Fig. 7.4, over the last 10 years, except in 2009 and 2014/2015 due to the U.S. triggered global financial crisis and China's stagnation respectively, all companies have grown, earning more than double against a backdrop of palm oil price increases. In addition, their ratio of pre-tax profit to revenue has been recorded at a level of more than 10% in good times at the mercy of palm oil price fluctuations. These huge gains serve as funds for accumulating wealth for executives, shown in Table 7.5, and for reinvesting the business expansion of vertical integration and global operations.

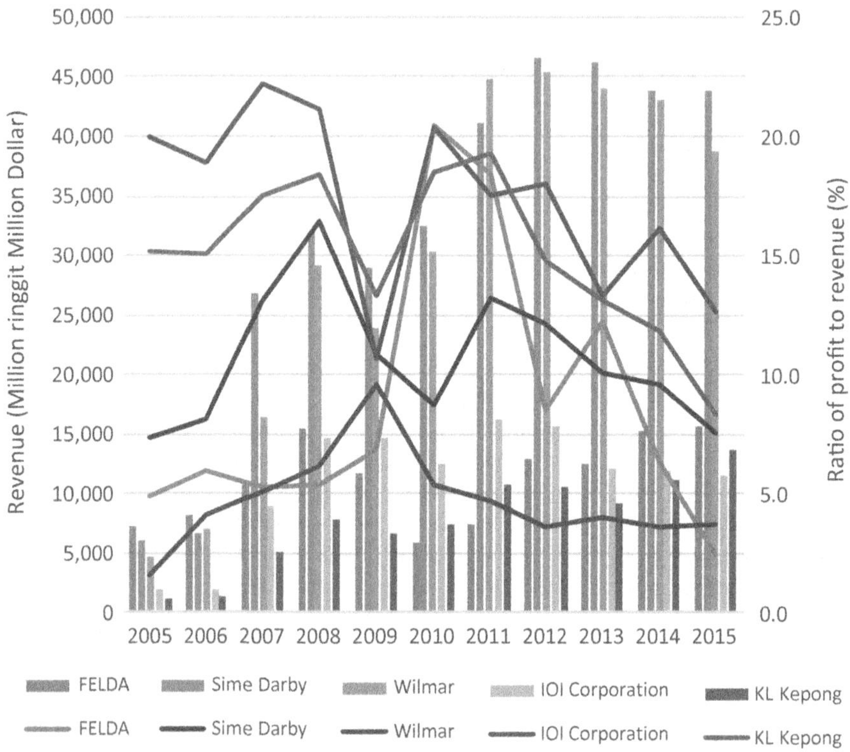

Fig. 7.4 Earning trends of leading agribusiness capital in Malaysia
Note: Bar graph indicates revenue, and line graph indicates profit before tax
Revenue is in dollar terms (Wilmar), in ringgit terms (other companies)
FELDA's data based on Felda Holdings until 2009. After 2010 based on Felda Global Ventures
Source: Prepared from each company's *Annual report,* various issues

Globalized Agribusiness and Its Impact on Palm Oil Industry

Extension of Overdevelopment and Sociopolitical Disruption

While the development of the Malaysian palm oil industry has generated the growth opportunity for agribusiness, recent globalized operations of agribusiness capital have brought about economic, sociopolitical, and ecological impacts in both host and home country.

First of all, extensive plantation development has broadly spread sociopolitical and environmental disruptions at local production sites. Along with limited land development, agribusiness capital has moved to invest in greenfield areas both domestically and abroad. In Sarawak, the last frontier of development in Malaysia, peatland forest clearing for oil palm plantations, which has caused climate change, reached more than 500,000 ha by 2007, over 38% of the total peatland surface area (Jalong 2012). And this transformation has also been accompanied by the land appropriation of Native Customary Right (NCR), which indigenous people have historically used for planting and hunting, without any discussion or prior consent. This has become a strong political issue, leading to protests, demonstrations, and transportation blockages by local people, which have been carried out repeatedly. According to the report by Forest Peoples Programme and Sawit Watch, about 40 lawsuits related to oil palm development by native plaintiffs occurred from 1997 to 2007 (Colchester et al 2007; Iwasa 2011). For example, a lawsuit between the Teran Kanan community and IOI has continued for 12 years and finally ruled in favor of the community in 2010. However, IOI has continued to occupy the related land site, and the Sarawak State Government also gave silent approval (Jalong 2012). Urgent solutions like Free, Prior, and Informed Consent and people's participation in assessment are still necessary. Indonesia has also faced similar issues such as large-scale biodiversity loss, massive deforestation, illegal logging, forest fires, land dispute, and human violation after the oil palm development boom, in which Malaysian agribusinesses have been involved. In the case of first transnational pollution caused by haze in 1997, 176 companies were accused of forest fire, of which 133 were oil palm-related companies, including 43 Malaysian agribusinesses (Wakker 2005; Colchester et al. 2006). As Malaysian capital as well as Indonesian one pushed into Indonesia to invest for massive plantation development, these contradictions have also extended across borders, and consequently, palm oil has been exposed to international criticism. Especially, haze affecting not only Indonesia but also neighboring countries has created an "anti palm oil campaign" by international NGOs and the establishment of Sawit Watch, an Indonesian NGO for monitoring oil palm overdevelopment (Chao and Colchester 2012; Jiwan 2013).

In the process of increased tension between development and regulation, Roundtable on Sustainable Palm Oil (RSPO) was established in 2004 for the purpose of managing overdevelopment. RSPO is characterized by encouraging participation from all stakeholders of palm oil commodity chains to discuss the production and distribution of sustainable palm oil. RSPO members increased to 1631 in 72 countries by June 2014, the principles and criteria (P&C) for sustainable palm oil were defined in 2005, and a palm oil certification system was introduced in 2008. Using this framework has made some progress toward sustainable plantation development such as allegation made by NGOs of deforestation and peatland clearance by Sinar Mas/GAR and subsequent pressure on Unilever to cease procurement from the company (Wright 2009). Major agribusinesses like Wilmar have been forced to change their behavior and to announce a commitment to address a No Deforestation Policy (Maitar 2013).

However, the penetration of certified palm oil is still at a low level. The percentage of certified palm oil production is 18%, and certified distribution volume is only half of the certified production (RSPO 2014). In addition, conflicts of interest between plantation industry groups and NGOs are also a barrier to spread the certified palm oil system, and, in contrast to RSPO, Indonesia and Malaysia have launched more producer-oriented certification schemes with no stakeholder/NGO participation and weaker standards in terms of the recognition of customary rights and Free, Prior, and Informed Consent for communities (Down to Earth 2011). Forest fires and transnational haze caused by slash-and-burn methods for oil palm and pulp plantations still occur annually and extend damages to people's livelihood across borders (The Straits Times 2015). Recently, after major buyers' ceased sourcing, IOI sued RSPO and suspended its sustainability certification over deforestation allegations in Malaysian Borneo and West Kalimantan on May 2016. However, they withdrew their legal challenge and announced the introduction of an action plan in line with RSPO's highest level of accreditation by the end of 2016 (Vaugham 2016; Mathiesen 2016).

Furthermore, there are growing concerns about global expansion of plantation development called "land grabbing" (White et al. 2013). Malaysian companies have already developed one million ha in Indonesia and are predicted to increase to 2 million ha by 2020, adding RM17.6 billion to GNI growth in Malaysia (MPOB and Ministry of Plantation Industries and Commodities 2011). According to the Land Matrix, a global and independent land monitoring initiative, Malaysia is now the world's second-largest investor behind the U.S. in line with overseas land investment (Land Matrix 2015). As noted above, agribusinesses like Sime Darby and Wilmar are expanding to invest in Africa in search of vast tracts of land and in response to the restriction to the overdevelopment in Indonesia, the present principal production areas. This pattern of development has reproduced violation of land rights of local communities and crisis of people's livelihood (Silas 2012; GRAIN 2013). In other words, such cross-border development is bringing out a new politics of "enclosure" in the country presently facing a food crisis and is subject to new global scrutiny.

International Competition and Commodity Chain Restructuring

Second, as agribusinesses tend to become globalized, linkages between domestic palm oil commodity chains are weakening. While palm oil downstream sectors have progressed in conjunction with the upstream plantation sector, the recent slowdown of the plantation sector due to land and labor problems has raised sluggish supply of domestic crude palm oil and over-capability of the refinery sector. As a solution to ensuring a stable supply, Malaysian refiners have increased the import of crude palm oil from Indonesia.

Along with Indonesia gaining power as a prime producer, however, international competition between Indonesia and Malaysia has become more intense. Previously, Indonesia has exported palm oil mainly in the crude form, albeit based on an abundant supply. Yet, Indonesia has gradually promoted industrial policies encouraging domestic palm oil refining for export. A symbolic case is the export duty change implemented by the Indonesian government in September 2011. With this measure, through tariff reductions to refine oil in Indonesia, local crude palm oil became cheaper than in Malaysia, and, as a result, refining companies in Malaysia now face a competitive disadvantage for exporting processed palm oil (Ng 2012).

On the other hand, the Malaysian government has instituted an export quota system of crude palm oil. This quota is allocated to agribusinesses, which hold refinery plants overseas such as Sime Darby, FELDA, and IOI. In other words, this quota system serves measures suitable for a transnational vertical integration strategy by producing FFBs in their own plantations, exporting crude oil and processing in their own overseas plants. However, contrary to the benefit of transnational agribusinesses, this quota system has been disadvantageous to refinery companies by suppressing domestic crude oil supply and decreasing the low factory utilization and profitability (New Straits Times 2012).

Under international competition and difficult domestic procurement, refinery business circles have put forward the plight of the Malaysian refining sector, especially nonintegrated refiners which do not hold plantations. PORAM (The Palm Oil Refiners' Association of Malaysia*)* has been demanding the abolition of the export quota system, which harms crude oil supply, and a revision of the Malaysian export duty (Interview with PORAM, August 13, 2012). The Malaysian government eventually announced a policy to change export taxes on CPO from 22 to 4.5–8.5% and to abolish the export quota from January 2013, leading to improve a tax gap between CPO and PPO to 5.5% and to decrease a price gap to U.S. $10–20. However, the Indonesian government, in turn, took a countermeasure against Malaysia to slash export taxes. Since then, international competition between Malaysia and Indonesia for palm oil export taxes has increased (Kagaku Kogyo Nippo 2014).

There are also concerns about an investment shift and deindustrialization of the refining sector. Indonesian policy to boost the downstream sector, coupled with low profitability in Malaysia, has caused a rush of investment by foreign agribusinesses such as Sime Darby and Wilmar as well as Indonesian capital from conglomerates such as Sinar Mas/GAR. As a result, refinery capacity in Indonesia eventually surpassed that of Malaysia to reach the total capacity of 30 million tons, and the Malaysian refinery business has been increasingly losing ground (Yushi 2014). As such, in contrast to the growth of globalized agribusinesses, the Malaysian palm oil industry linking upstream and downstream sectors nationally is now at the crossroads.

Import of Foreign Labor Force and Agricultural Vulnerability

Third, the transformation of the plantation labor force and agricultural vulnerability has escalated. Since the 1980s, plantation worksites have been undergoing a severe reduction and an aging of the labor force, due to outmigration to urban areas to avoid poor working/living conditions. On the other hand, labor processes in plantations such as harvesting, field management, transportation, etc. are labor-intensive and difficult to mechanize, and securing sufficient workers is, therefore, crucial to the survival of the industry.

A solution to this problem is to import a foreign labor force as another globalization strategy of agribusiness. According to the government's official figures, foreign workers in oil palm plantations are mostly from Indonesia accounting for nearly 90%, as well as a small number from Bangladesh, the Philippines, Nepal, and other countries. As for the job categories, foreign workers are mostly engaged in fieldwork such as harvesting, fruit collecting, and other manual labor, while Malaysians are mainly *mandors* (supervisor cum labor-recruiter) and executives staff. As a result, in 2012 foreign workers amounted to approximately 370,000 persons, accounting for three-quarters of the total labor force. In particular, they account for over 90% of harvesters, and are also included as part of management staff. Nevertheless, severe labor shortages still continue, the amount is 34,000 persons, little more than 6% of the total workforce requirement (Table 7.6).

The problem of accompanying dependence on foreign labor has been the vulnerability of the domestic agricultural production force. According to MPOB (*Malaysian Palm Oil Board*) survey, the main disadvantages of employing foreign labor force include labor recruitment, employment procedures, and high labor turnover. In addition to this, Indonesia, the main source of transnational labor supply, has been undergoing plantation expansion and related job growth, which makes Malaysian agribusinesses face difficulties recruiting field workers. From an Indonesian worker's point of view, it is unattractive to go across the border because payment in Malaysia is similar to that of Kalimantan, and there is a preference for jobs in Indonesia. Similarly, workers are gradually unwilling to return to the field after the Ramadan season, and sudden labor shortages can arise every 2 weeks to one month (Mohd 2010).

On the other hand, agribusiness capital has still appealed for the need of foreign workers. With "crisis consciousness" of regulating their import, MPOA (Malaysian Palm Oil Association), a business group of the plantation sector, asserted continuous employment of foreign workers, stating that "palm oil export earnings will decrease by RM10 billion if foreign workers decrease by 30 percent" (Mamat 2010). The state of Sarawak, today's production center, was also reported to have expressed concern over a labor shortage for harvest, and to address this problem through the shift of sourcing labor from Bangladesh rather than Indonesia (Sibon 2014). Recently, an expose of Bangladeshi workers' abuse in FELDA plantations drew controversy, and the RSPO Complaints Panel decided to revoke its certification for the concerned FELDA's area (Zain Al-Mahmood 2015; RSPO 2016).

It should be considered, however, that the persistent poor working/living conditions of plantation workers are the root of this labor shortage. According to the government, plantation workers' monthly income is estimated RM847-1017 (U.S. $243–292), while the price of crude palm oil is RM3000 ($861). The Minister of Plantation Industries & Commodities Tan Sri Bernard Dompok noted in the Parliament on October 2009 that the net monthly income of oil palm harvester was around RM1,700 ($488) and its wage level exceeded the poverty line of RM720 ($207). On the other hand, a plantation workers' support group criticized this figure

Table 7.6 Labor force composition in Malaysian oil palm plantations (1st quarter, 2012)

	Plantation workers (person)			Percentage by occupation (%)			Share of foreign workers (%)	Labor shortage	
	Total	Malaysians	Foreign workers	Total	Malaysians	Foreign workers		Actual number (person)	Rate of shortage (%)
Total	490,142	116,564	373,578	100.0	100.0	100.0	76.2	33,870	6.5
Field workers	454,264	82,054	372,210	92.7	70.4	99.6	81.9	33,261	6.8
Mandores	13,384	7933	5451	2.7	6.8	1.5	40.7	491	3.5
Harvesting mandores	14,238	5569	8669	2.9	4.8	2.3	60.9	775	5.2
Harvesters & Fruits collectors	195,616	14,082	181,534	39.9	12.1	48.6	92.8	16,265	7.7
Field workers	161,923	32,023	129,900	33.0	27.5	34.8	80.2	12,489	7.2
Others	69,103	22,447	46,656	14.1	19.3	12.5	67.5	3,241	4.5
Ofiice workers	35,878	34,510	1368	7.3	29.6	0.4	3.8	609	1.7
Managers	9,825	9677	148	2.0	8.3	0.0	1.5	158	1.6
Staffs	26,053	24,833	1220	5.3	21.3	0.3	4.7	451	1.7

Note: Corrected erroneous description
Source: MPOB (2012). Oil Palm Industry Performance 2011

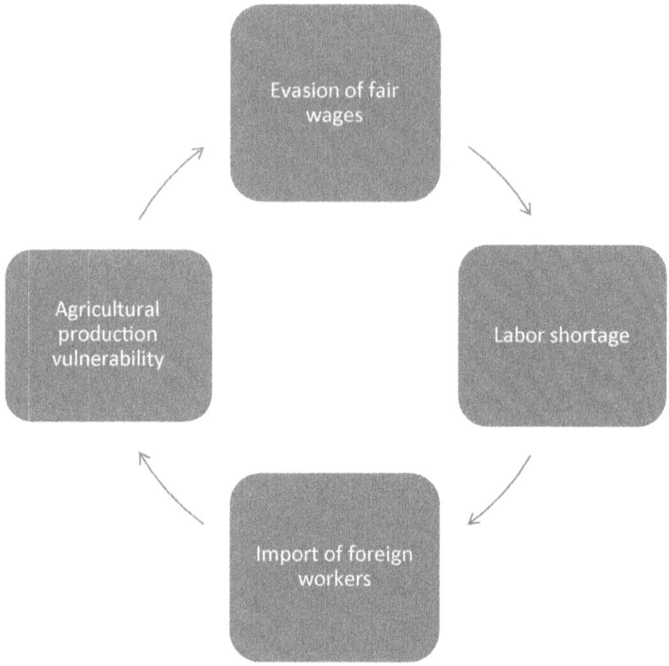

Fig. 7.5 Dependence on foreign workers and downward spiral

as not having calculated workers' risks to exposure to weather/soil conditions and not having included the increase of household expenses. According to their opinion, while agribusiness capital has made remarkable profits in recent years, workers' monthly income is less than RM650 ($187), in spite of calls for the basic income to be raised to RM750 ($215) since 1996, workers are still forced to live below the poverty line (Hance 2009; JERIT 2012).

This developing form of globalization and its impact thus neglects the improvement of domestic issues. The evasion of fair wages and a modern wage system has led to a domestic labor shortage, and agribusiness tends to deal with the situation through importing foreign workers. However, dependence on foreign labor creates a disincentive to improve workers' conditions, which leads to more labor shortages and deeper dependence on foreign labor. As Indonesian workers are not likely to cross the border due to an increase in employment opportunities in Indonesia, this spiral will deepen agricultural production vulnerability (Fig. 7.5).

Conclusion

Since the 1960s, Malaysia, as a pioneer of large-scale oil palm plantations development, has been the leading palm oil producer/exporter. In parallel to plantation development, palm oil related agro-industrialization has realized both a diversification and upgrading for market export. As a result, Malaysia has been one of the NACs to arise through its emerging palm oil exports. In recent years, however, the Malaysian palm oil industry has slowed down due to land and labor shortages, and lost its top production rating to Indonesia, which has led to increased international competition between the two countries.

In contrast to national industrial trajectory, Malaysian agribusiness capital has extended its penetration from the home country to a worldwide level. Based on plantation development, agribusiness capital has implemented strategies not only for vertical integration through mergers and acquisitions and strategic alliances with transnational corporations but also of transnationalization of both upstream and downstream sectors. Especially in the plantation sector, breaking through the restrictions of land and labor problems, agribusiness has extended its development space from East Malaysia to Indonesia on the one hand, and imported foreign labor force mainly from Indonesia on the other. Through these strategies, agribusiness has reaped considerable fruits from the palm oil business and has further used their gains to accumulate the wealth of executives and to reinvest business expansion globally.

However, recent cross-border operation of agribusiness capital has brought about economic, sociopolitical, and ecological contentions in both host and home country. In the host countries, extensive plantation development has accompanied sociopolitical and environmental disruption. In spite of the establishment of RSPO, its progress is slow and currently, new development issues such as transnational land grabbing, international air pollution caused by haze, and sociopolitical and environmental conflicts are arising. In addition, the Malaysian palm oil industry has also been facing difficulties. In the plantation sector, agribusiness has heavily depended on foreign labor to avoid the fair wages for domestic workers, and this creates a vicious cycle of heavy labor shortage and serious dependence on foreign labor, which leads to deepening agricultural vulnerability. Also in the downstream sector, under the difficult domestic procurement due to agricultural vulnerability and Indonesia's surge, the Malaysian refinery sector has faced a serious plight.

Today, in the context of the global palm oil demand on the one hand and national/international development monitoring on the other, major agribusinesses are moving to the next stage of expanding their investment space from Malaysia/Indonesia to outside Southeast Asia, which leads to organized palm oil commodity chains or "global palm connection." In contrast to the growth of globalized agribusinesses avoiding socio-environmental crisis through "spatial fixes" (Harvey 1982; p. 415), the Malaysian palm oil industry linking upstream and downstream sectors nationally is standing at a crossroads through restructuring national commodity chains before building a substantial national economic base that has the potential to create radical domestic and transnational change.

References

ADM. (2012, October 18). ADM and Wilmar receive approval for partnerships in fertilizer, ocean freight and vegetable oil. *ADM press release.* Retrieved October 20, 2012, from http://www.adm.com/news/_layouts/PressReleaseDetail.aspx?ID=446.

Adnan, H. (2017, September 16). IOI Corp and Bunge in win-win deal. *The Star.* Retrieved September 20, 2017, from https://www.thestar.com.my/business/business-news/2017/09/16/ioi-corp-and-bunge-in-winwin-deal/.

Bair, J. (Ed.). (2009). *Frontiers of commodity chain research.* Stanford: Stanford University Press.

Borneo Post. (2013, July 2). Felda global terminates JV with Canadian Bunge. *Borneo Post.* Retrieved July 3, 2013, from http://www.theborneopost.com/2013/07/02/felda-global-terminates-jv-with-canadian-bunge/.

Chao, S. & Colchester, M. (Eds.). (2012). *Human rights and agribusiness: Plural legal approaches to conflict resolution, institutional strengthening and legal reform.* Moreton-in-Marsh: Forest Peoples Programme & Perkumpulan SawitWatch.

Colchester, M., Jiwan, N., Andiko, S. M., Firdaus, A. Y., Surambo A., & Pane, H. (2006). *Promised land: Palm oil and land acquisition in Indonesia.* Moreton-in-Marsh: Forest Peoples Programme, Perkumpulan Sawit Watch, HuMA & World Agroforestry Centre.

Colchester, M., Pang, W. A., Chuo, W. M., & Jalong, T. (2007). *Land is life: Land rights and oil palm development in Sarawak.* Moreton-in-Marsh: Forest Peoples Programme & Perkumpulan Sawit Watch.

Department of Statistics Malaysia. (2011). *Population and housing census of Malaysia 2010: Population distribution and basic demographic characteristics.* Putrajaya: Department of Statistics Malaysia.

Department of Statistics Malaysia. (2012). *Banci ekonomi 2011 manufacturing.* Putrajaya: Department of Statistics Malaysia.

Department of Statistics Malaysia. (2015). *Malaysia economics statistics – Time series: External Trade.* Department of statistics Malaysia. Retrieved October 16, 2017, from https://www.statistics.gov.my/index.php?r=column/ctimeseries&menu_id=NHJlaGc2Rlg4ZXlGTjh1SU1kaWY5UT09.

Down to Earth. (2011, April). Indonesian sustainable palm oil scheme to speed up palm oil development. *DTE*, 88.

Felda Global Ventures Holdings Berhad. (2016). *Empowering sustainable value: Annual integrated report 2015.* Kuala Lumpur: Felda Global Ventures Holdings Berhad.

Fold, N., & Prichard, B. (Eds.). (2005). *Cross-continental food chains.* London: Routledge.

Foster, J. B. (2000). *Marx's ecology: Materialism and nature.* New York: Monthly Review Press.

Gereffi, G., & Korzeniewicz, M. (Eds.). (1994). *Commodity chains and global capitalism.* Westport: Greenwood Press.

GRAIN (2013, December 7). Stolen land: Nigerian villagers want their land back from Wilmar. *GRAIN website.* Retrieved December 8, 2013, from https://www.grain.org/article/entries/4844-stolen-land-nigerian-villagers-want-their-land-back-from-wilmar.

Hance, J. (2009, November 19). Oil palm workers still below poverty line, despite minister's statements. *Mongabay.com.* Retrieved December 1, 2009, from http://news.mongabay.com/2009/1119-hance_oilpalmworkers.html.

Harvey, D. (1982). *The limits to capital.* Oxford: Basil Blackwell.

Iwasa, K. (2005). *Mare-sia ni okeru nougyou kaihatu to aguribijinesu: Yusyutsu-shikougata kaihatsu no hikari to kage* [Agricultural development and agribusiness in Malaysia: The Light and shadow of export-led development]. Kyoto: Horitsu-bunka-sya Pub. (in Japanese).

Iwasa, K. (2011). Limits to agribusiness-led development: A structural analysis of the Malaysian palm oil industry. *Kochi University Review of Social Sciences, 102,* 1–19.

Jalong, T. (2012). Experiences with oil palm expansion in Sarawak: the need for new standards. In S. Chao & M. Colchester (Eds.). *Human rights and agribusiness: Plural legal approaches to conflict resolution, institutional strengthening and legal reform.* Moreton-in-Marsh: Forest Peoples Program & Perkumpulan SawitWatch.

JERIT (Jaringan Rakyat Tertindas). (2012, July 3). Sime Darby gets increased profits! Workers still get colonial wages. *Press statement.*

Jiwan, N. (2013). The political ecology of the Indonesian palm oil industry. In O. Pye & J. Bhattacharya (Eds.). *The palm oil controversy in Southeast Asia: A transnational perspective* (pp. 48–75). Singapore: Institute of Southeast Asian Studies.

Kagaku Kogyo Nippo. (2014, November 11). *Pa-mu yu Tonan Ajia de yusyutu kyoso gekika yusyutsu kanzei zero eikyo* [Intensifying palm oil export competition in Southeast Asia: The effects of zero export tax]. *Kagaku Kogyo Nippo* [The Chemical Daily] (in Japanese).

Khera, H. S. (1976). *The oil palm industry of Malaysia: An economic study.* Kuala Lumpur: Penerbit Universiti Malaya.

Land Matrix. (2015). *Web of transnational deals.* Land Matrix website. Retrieved January 13, 2015, from http://www.landmatrix.org/.

Levitt, T. (2011, March 25). Palm oil giants target Africa in 'land grab' following Indonesia deforestation ban. *The Ecologist.*

Maitar, B. (2013, December 6). You did it! Palm oil giant commits to no deforestation. *Greenpeace International website.* Retrieved March 10, 2014, from http://www.greenpeace.org/international/en/news/Blogs/makingwaves/palm-oil-giant-Wilmar-commits-no-deforestation/blog/47623/.

Mamat, Dato' S. (2010). Foreign labour: Issues & challenges (MPOA's Views). In *Proceedings of the palm industry labour: Issues, performance & sustainability (PILIPS) Workshop.* Kelana Jaya: MPOB.

Marx, K., & Fowkes, B. (Trans.). (1867/1977). *Capital: A critique of political economy* (Vol. 1). New York: Vintage Books.

Mathiesen, K. (2016, June 7). Malaysian palm oil giant IOI drops lawsuit against green group. *The Guardian.* Retrieved August 8, 2016, from https://www.theguardian.com/sustainable-business/2016/jun/07/palm-oil-ioi-rspo-unilever-nestle-kelloggs-mars-deforestation-indonesia.

McCarthy, J., Gillespie, P., & Zahari, Z. (2012). Swimming upstream: Local Indonesian production network in 'globalized' palm oil network. *World Development, 40*(3).

McMichael, P. (1996). *Development and social change.* Thousand Oaks: Pine Forge Press.

Mohd, A. S. (2010). Economic impacts of foreign labour. In *Proceedings of the palm industry labour: Issues, performance & sustainability (PILIPS) Workshop.* Kelana Jaya: MPOB.

MPOB and Ministry of Plantation Industries and Commodities. (2011). *NKEA palm oil: The way forward.* Kelana Jaya: MPOB.

MPOB (Malaysian Palm Oil Board). (2014). *Malaysian oil palm statistics 2013.* Kelana Jaya: MPOB.

New Straits Times. (2012, August 1). There's a lot at stake, say palm oil refiners. *New Straits Times.*

Ng, L. F. (2012, April 6). Malaysian refiners' woes. *CIMB Report.*

PEMANDU (Performance Management & Delivery Unit). (2010). *Economic transformation programme: A roadmap for Malaysia.* Putrajaya: PEMANDU.

Pye, O. (2013). An analysis of transnational environmental campaigning. In O. Pye & J. Bhattacharya (Eds.). *The palm oil controversy in Southeast Asia: A transnational perspective* (pp. 179–198). Singapore: Institute of Southeast Asian Studies.

RSPO (Roundtable on Sustainable Palm Oil). (2014). *Impact report 2014*. Kuala Lumpur: Roundtable on Sustainable Palm Oil. Retrieved May 10, 2016, from http://www.rspo.org/file/14_0082RoundtableonSustainablePalmOil(RSPO)ImpactReport2014v14-spread.pdf.

RSPO (Roundtable on Sustainable Palm Oil). (2016). Status of complaints: Case tracker, FELDA (Felda Global Ventures). *RSPO website*. Retrieved May 26, 2016, from http://www.rspo.org/members/complaints/status-of-complaints.

Sibon. P. (2014, April 6). Harvest is great, workers are few. *The Borneo Post*.

Silas, K. S. (2012). *Uncertain futures: The impacts of Sime Darby on communities in Liberia*. Montevideo: World Rainforest Movement & Sustainable Development Institute.

Sime Darby Plantation Sdn Bhd. (n.d.). Sime Darby in Liberia: The new way forward. Retrieved January 10, 2015, from http://www.simedarbyplantation.com/Sime_Darby_in_Liberia-;_The_New_Way_Forward.aspx.

Singh, G., Lim, K. H., Teo, L., & Chan, K. W. (2009). *Sustainable production of palm oil: A Malaysian perspective*. Kuala Lumpur: Malaysian Palm Oil Association.

Tan, C. K. (2014, July 18). Malaysia's Felda Global aims for eightfold sales rise by 2020. *Nikkei Asian Review*.

Tan, P. L. (2008). *Land to till: The Chinese in the agricultural economy of Malaysia*. Kuala Lumpur: Centre for Malaysian Chinese Studies.

Teoh C. H. (2013). Malaysian corporations as strategic players in Southeast Asia's palm oil industry. In O. Pye & J. Bhattacharya (Eds.). *The palm oil controversy in Southeast Asia: A transnational perspective*. Singapore: Institute of Southeast Asian Studies.

The Straits Times. (2015, December 18). Efforts to stop Indonesian haze fires may not work for 2016. *The Straits Times*. Retrieved December 19, 2015, from http://www.straitstimes.com/asia/se-asia/efforts-to-stop-indonesian-haze-fires-may-not-work-for-2016.

UNCTAD (United Nations Conference on Trade and Development). (2009). *World investment report 2009: Transnational corporations, agricultural production and development*. New York: United Nations.

Vaugham, A. (2016, May 9). Top palm oil producer sues green group over deforestation allegations. *The Guardian*. Retrieved May 9, 2016, from https://www.theguardian.com/environment/2016/may/09/top-palm-oil-producer-sues-green-group-over-deforestation-allegations.

Wakker, E. (2005). *Greasy palms: The social and ecological impacts of large-scale oil palm plantation development in Southeast Asia*. London: Friends of the Earth.

Wright, T. (2009, December 13). Unilever drops palm oil supplier. *The Wall Street Journal*.

White, B., Borras, S., Jr., Hall, R., Scoones, I., & Wolford, W. (Eds.). (2013). *The new enclosures: Critical perspectives on corporate land deals*. London: Routledge.

Yantoultra, N., & Niluksi, K. (2012, June 28). Malaysia's Felda Surges 20 percent in debut of World's No.2 IPO. *Reuters*. Retrieved June 30, 2012, from https://www.reuters.com/article/us-malaysia-felda-ipo/malaysias-felda-surges-20-percent-in-debut-of-worlds-no-2-ipo-idUSBRE85R07S20120628.

Yushi. (2014). *Jun-cho na seisan to yusyutsu fushin de jukyuu yurumu* [Favourable production and decline in exports caused positive supply-demand conditions]. *Yushi*, 67(7), 44–46 (in Japanese).

Yusof, B. (2008). Malaysia's oil palm: Hallmark of sustainable development. *Global Oils & Fats Business Magazines*, 5(4), 1–7.

Zain Al-Mahmood, S. (2015, July 26). Palm oil migrant workers tell of abuses on Malaysian plantations. *The Wall Street Journal*. Retrieved July 29, 2016, from http://www.wsj.com/articles/palm-oil-migrant-workers-tell-of-abuses-on-malaysian-plantations-1437933321.

Chapter 8
Honey Bees in Modernized South East Asia: Adaptation or Extinction?

Panuwan Chantawannakul

Abstract Honey bees (*Apis* spp.)—key pollinators—have evolved over millions of years along with flowering plants and Southeast Asia is a key biodiversity hotspot for honey bees. They are known to play a crucial role in maintaining plant biodiversity in forests and agricultural crops for human food production. This chapter describes the history of beekeeping, beekeeping practices, and offers an overview of the impact of changing environments as well as human activities on honey bee populations in Southeast Asia. It presents an overview of the challenges that honey bees currently face in Southeast Asia and contextualizes traditional knowledge and beekeeping practices to harvest bee products that have long been embedded in local cultures. The chapter also examines and introduces how, in recent decades, societies in Southeast Asia have undergone significant transformations, which have severely affected wild and domesticated honey bee populations. It shows how under these conditions some species have adapted to the new environment, whereas other bee populations have been rapidly decreasing. A combination of human activities (e.g., land use and deforestation), newly emerging diseases and climate change have meant that they are struggling to adapt and maintain species survival in the region.

Keywords Asian honey bees · Deforestation · Conservation · Land use
Bee health · Pollinators · Bee diseases · Bee parasites · Biodiversity

Introduction

Honey bees are social insects that not only produce food but also play an important part in religious beliefs and cultures. The history of Asian honey bees is extremely ancient and there are numerous depictions of honey being offered to Buddha, as can be seen in temples in Thailand (Fig. 8.1). Within Southeast Asia region, bees have

P. Chantawannakul (✉)
Department of Biology, Faculty of Science, Chiang Mai University,
50200 Chiang Mai, Thailand
e-mail: panuwan@gmail.com

M. Lopez and J. Suryomenggolo (eds.), *Environmental Resources Use and Challenges in Contemporary Southeast Asia*, Asia in Transition 7,
https://doi.org/10.1007/978-981-10-8881-0_8

Fig. 8.1 A monkey offering a harvested colony of *A. florea* to the Buddha. This carving depicts an occasion when the Buddha went to the jungle for mediation relating the story of when the elephant and monkey brought him food, which were bananas and dwarf honey bee combs. Photo taken by author at Wat Jed Yod temple, Chiang Mai, Thailand

also been mentioned in legends. These have served as elements of honey-hunting practices in the region and offer us insights into the rich traditions that have arisen out of human-ecological interactions (Oldroyd and Wongsiri 2006).

Honey has also been recorded as an important food in Islamic, Buddhist, and Hindu religious texts (Chantawannakul et al. 2004; Oldroyd and Wongsiri 2006). In Thai Buddhism, monks are permitted to consume honey as a tonic in the evenings, while only consuming one or two meals a day before noon. For instance, during *Vassa* or Buddhist Lent in Laos, Myanmar, Cambodia, and Thailand, a three-month annual retreat in the rainy season, monks are confined to their temples and beeswax is used to make candles for monks during their meditation (Kobayashi 2013). Honey is also frequently mixed with foods and shared communally during Thai Buddhist festivals. Furthermore, as within other world regions, honey has also been used as a natural sweetener in Southeast Asia. Uses in the region range from topical applications for burns and abrasion, inclusion in traditional herbal medicines, and treating stomach and digestive complaints (Wongsiri et al. 1995; Chantawannakul et al. 2004), and recently, the biomedical activities of Thai honey have been confirmed scientifically. Different antibacterial and antioxidant activities, as well as physiochemical properties of honey are dependent on their nectar sources, such as longan, lychee, coffee, sunflower, sesame, bitter bush, and para-rubber (Wanjai et al. 2012; Pattamayutanon et al. 2015). In northern Thailand, honey is considered to be an elixir if consumed daily with bananas (Chantawannakul et al. 2004). Researchers have noted that humans have hunted honey bees and eaten both honey and broods for feral colonies of honey bees (Crane 1999). There are several traditional recipes using bee brood and honey for local cuisine. For example, in Thailand they are cooked with chicken eggs, which can be made by using a bee brood comb wrapped in banana leaves and steamed in Thailand (Chen et al. 1998). Eating insects are more common in rural rather than urban areas. However, increasing urbanization in the region has meant that entomophagy is in decline (van Huis et al. 2013). Other bee products such as propolis (a plant resin collected by bees), royal jelly, and bee pollen also show various levels of consumption. Propolis has become popular for medical uses as it has been proven to possess antimicrobial and antioxidant compounds (Sanpa et al. 2015; Bankova et al. 2000). Thai propolis, in particular, was found to have bioactive compounds from mangoes and mangosteens (Sanpa et al. 2016).

Most importantly for Southeast Asia, honey bees are the main pollinators (Stout and Morales 2009), whose pollination services have been markedly undervalued since they are more known for producing honey rather than for their role as crucial pollinators. At present, it has been estimated that global crop production (approximately 35%) depends on insect pollination (Klein et al. 2007) with more than 70% of the leading food crops worldwide dependent upon insect pollination. As such, there is a dearth of information on the role of honey bee pollination in Southeast Asia.

This chapter will discuss the use of honey bees and beekeeping in Southeast Asia, with special reference to modern beekeeping techniques that were introduced to the region in the late twentieth century. It focuses on environmental changes affecting the region, potential honey bee pathogens, and parasites that have affected honey bees and the challenges of beekeeping as an industry for local communities. Based on this discussion, this chapter argues that the beekeeping industry must develop a positive holistic approach, which includes input from all stakeholders: scientists, policymakers, beekeepers, and business partners, to manage environmental costs and also present and future conservation of honey bee species in Southeast Asia.

Evolution and Diversity of Honey Bees in Southeast Asia

Honey bees have been evolving for more than thirty million years. The first fossil species of the genus discovered in Europe was *Apis henshawi* (Cockerell 1907, 1909). Another bee fossil *Apis javana* (Stauffer 1979) of the Cenozoic era was found in Malaysia. This species is believed to be a close ancestor of *Apis cerana* (Fabricius 1793). Approximately every million years 20–25% of all species are replaced, numerous extinctions and other speciation events take place (Rosenzweig 1995), and nine extinctions have occurred during the Miocene and Oligocene of the Cenozoic era, over 41 million years ago (Engel 1998; Spellerberg and Sawyer 1999) (i.e., *Apis armbrusteri* (Zeuner and Manning 1976), *Apis catanensis* (Roussy 1960), *Apis cuenoti* (Théobald 1937), *Apis henshawi* (Cockerell 1909), *Apis longtibia* (Zhang 1990), *Apis melisuga* (Zeuner and Manning 1976), *Apis miocenica* (Zhang 1990), *Apis petrefacta*, and *Apis vetustus* (Engel 1998)). Approximately two million years ago, there were at least nine extant species present since the Holocene and Pleistocene eras (Engel 1998; Spellerberg and Sawyer 1999): *Apis andreniformis* (Smith 1858), *A. cerana* (Fabricius 1793), *Apis dorsata* (Fabricius 1793), *Apis laborisa* (Smith 1871), *Apis koschevnikovi* (Tingsek et al. 1988), *Apis florea* (Fabricius 1787), *Apis nigrocincta* (Smith 1861), *Apis nuluensis* (Tingsek et al. 1996), and *A. mellifera* (Linnaeus 1758). In the modern period, through translocation by humans and economic activities, *Apis mellifera* has become a cosmopolitan species, whereas the other eight extant species have maintained ecological niches in Asia.

Southeast Asia has the largest diversity of honey bees covering all nine native species and one introduced species (*A. mellifera*), which has now become widespread in Southeast Asia (Fig. 8.2). All are regarded as one of the key pollinators that maintain forest ecosystems in the region and are grouped according to their nesting structures (i.e., open and cavity nesting) (Rahman and Rahman 2000; Buawangpong et al. 2014). Cavity nesting Asian honey bees (i.e., *A. cerana*, *A. koschevnikovi*, *A. nigrocincta*, and *A. nuluensis*) are commonly found in protective cavities in small caves or tree hollows honeybees (Oldroyd and Wongsiri 2006). *A. cerana* is more cosmopolitan, compared to the other native species and distributed throughout Southeast Asian countries. The other three Asian cavity nesting honey bees are restricted to other areas in Southeast Asia (Fig. 8.2). The giant honey bee (*A. dorsata, A. laboriosa*) that frequently hang their open air nests on the branches of tall trees and cliffs, nowadays have adapted themselves to nest on high concrete buildings (Wongsiri et al. 1996) and are geographically limited to warm climates (Crane 2003). The two dwarf Asian honey bee species (*A. florea* and *A. andreniformis*) have open air nests. *A. florea* is also found in Southeast countries (Crane 2003) and recently, there is evidence that supports the recognition of the giant Philippines honey bee *Apis breviligula* as a separate species from *A. dorsata* (Lo et al. 2010). In the 1990s, the European honey bee (*A. mellifera*), which can be domesticated, was introduced into the region. The species diversity of bees is

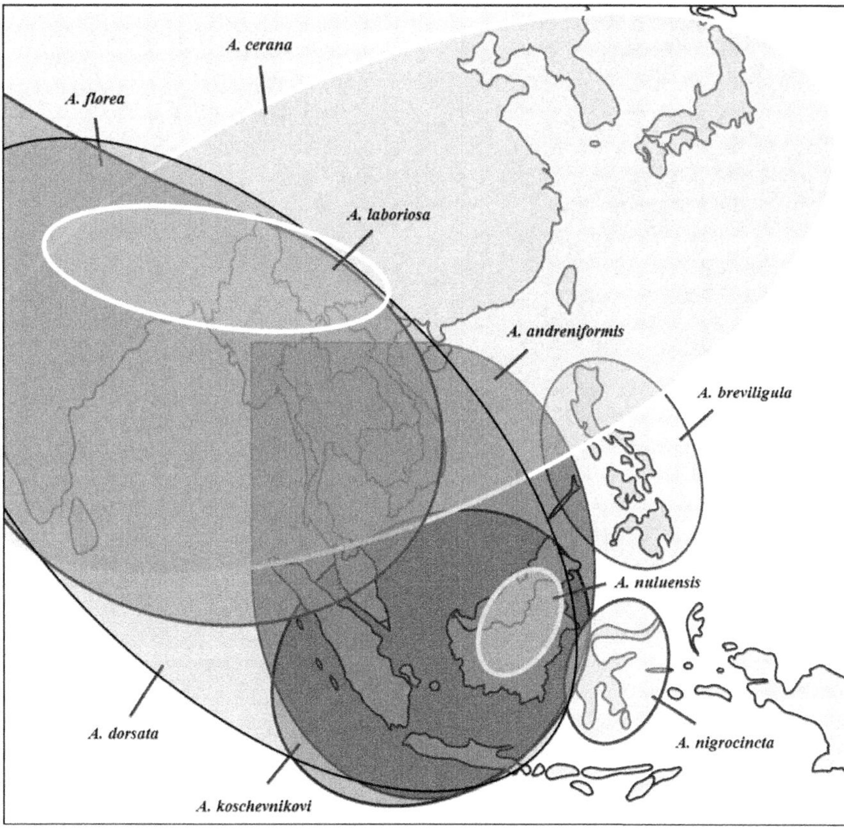

Fig. 8.2 Distribution map of Asian honey **bee species in Southeast Asia**
Source Author

highest in the tropics and decrease toward the poles (Spellerberg and Sawyer 1999). This pattern of honey bee distribution holds the key to the survival of honey bees.

Hunting Wild Bees

Southeast Asian societies have commonly exploited wild bees through a honey-gatherer lifestyle near forested areas. Local people in Southeast Asia still maintain gathering practices to exploit wild honey bees (Chantawannakul et al. 2004). *A. dorsata* and *A. florea* are opportunistically harvested or actively sought out by honey hunters (Fig. 8.3a, b). Harvesting wild honey from the giant honey bee is usually risky and a destructive operation as it requires cutting the combs (Oldroyd and Wongsiri 2006). The harvesting period in Northern Thailand falls during February–April annually, which is a dry season and where the moisture in the

Fig. 8.3 **a** Dwarf honey bee (*A. florea*). **b** Giant honey bee (*A. dorsata*). The photos were taken in Chiang Mai, Thailand. Photo by the author

honey is low and less likely to be fermented. Bee hunters observe the bees from the time of their arrival at the tree and often choose hives which are full of honey. In particular, hunters harvest honey immediately prior to the colony's migration which can be observed through bee behavior such as when the queen's brood laying rate is reduced (Saraithong et al. 2012). Such operations can minimize brood loss. Hunters prefer to take down the beehives at night, especially moonless ones, or early mornings. In Thai local cultures, it is also believed that bee sites or bee trees have angels living in them and protecting the trees. Before the operation, some hunters will pray or chant for permission and for a safe and successful operation. They use bamboo ladders to climb trees or rope wrapped around the trunk and climb the trees and light their torch and brush the comb. In peninsular Malaysia and Sumatra, for example, hunters often strike the branch with the torch to cause sparks and chant for the bees to "follow the stars" (Buchmann and Nabhan 1996; Oldroyd and Wongsiri 2006). When most of the adult bees are brushed off the comb, the hunter will cut off the comb having a basket hanging under it. In some parts of Southeast Asia, only honey is consumed, but in many other areas of the region, both honey and the brood are consumed, e.g., Thailand (Oldroyd and Wongsiri 2006). Honey is filtered with white cotton and the brood is cut into small squares. In Vietnam, Borneo, Sumatra and Sulawesi, Malaysia, and Cambodia, rafter beekeeping[1] of *A. dorsata* is also carried out (Hadisoesilo 2000; Waring and Jump 2004). Honey collectors take their harvests just before the colonies are due to migrate. In Southeast Asian countries, where rafter beekeeping is practiced, this is the most practical way of sustainable harvesting of *A. dorsata*. One rafter can yield approximately four kgs of harvested honey and they can potentially produce up to several harvests per year (Oldroyd and Wongsiri 2006). The wild honey from the giant honey costs about US $5 dollar per kg (personal observation; Oldroyd and Wongsiri 2006)

Hunting dwarf honey bees (*A. florea*) is relatively easy, since they nest in lower sites such as tree branches in orchards and are not aggressive. Hunters shake nests vigorously and once the queen is airborne the adult worker bees will follow and leave the comb. Hunters often cut the supporting branch and bind it into a bamboo tripod. In Thailand, 40,000–50,000 nests are sold annually in local markets (Chen et al. 1998; Oldroyd and Wongsiri 2006). At present, wild honey is still popular as it is perceived to be a pure and healthy product in Thailand (Chantawannakul et al. 2004; Oldroyd and Wongsiri 2006).

Traditional honey bee hunting and beekeeping are important in the spiritual and economic life of Southeast Asia. However, beekeeping practices are less well developed in the region than in temperate zones. Only one species has been domesticated in the region, where the open-nesting species are hunted for their honey. The availability of wild bees that can potentially be easily hunted is a key factor in harvesting honey.

---

[1]Rafter beekeeping is where hunters place tree-poles in the ground at a slight angle creating an environment for honeybees to migrate and populate.

As Southeast Asia is abundant with wild bees it is not necessary to domesticate cavity-nesting bees as can be seen in other parts of the world (Chantawanankul et al. 2016). Societies in the region also use other sweeteners that are derived from other crops, such as sugarcane, sweet sap from coconuts, and palm sugar. However, many local Southeast Asians build hives to attract bee swarms. Bait hives vary from the hollow log hives deployed in either a vertical or horizontal position, or simple man-made containers built of local materials that are available such as clay pots and straw baskets (Chantawannakul et al. 2004). Honey can be harvested by opening hives and cutting out the honey combs. The cavity nesting Asian honey bee, *A. cerana*, has been kept through a traditional approach for a long time to produce honey and wax for trade items. The Tai, ancestors of the modern Thai-Lao-Shan ethnolinguistic group, have practiced beekeeping with *A. cerana* in Xishuang Banna of the southern Yunnan Province in China for more than one thousand years (Wongsiri et al. 1987). Nevertheless, beekeeping in hollow log hives is not widely practiced in Thailand, Malaysia, and the Philippines. In Thailand, farmers often keep them with their agricultural areas such as the coffee plantations in Doi Saket, Chiang Mai, and they can harvest about 2.4 kg/colony/year, which is much lower than that of the European honey bee (35–40 kg/colony/year) (Fig. 8.4). Wild honey are also commonly harvested from *A. cerana*, *A. andreniformis*, *A. nuluensis,* and *A. koschevnikovi* in Malaysia. They all are proven to contain high polyphenol contributing to their antimicrobial activities (Yap et al. 2015)

Fig. 8.4 *Apis cerana* beekeeping in Chiang Mai, Thailand. Photo taken by the author

Modern Beekeeping Practices

Modern beekeeping management (keeping bees in a standard bee box) with European honey bees (Fig. 8.5) was introduced to Southeast Asia by European and American missionaries in the early 1990s (Oldroyd and Wongsiri 2006). Since then, modern beekeeping has expanded to many parts of the region, for example, *A. mellifera* was first introduced to Thailand for research purpose at Chulalongkorn University, Bangkok in the early 1940s and successfully expand to other areas in 1975 and now Thailand has more than 200,000 colonies (Oldroyd and Wongsiri 2006; Sanpa and Chantawannakul 2009). Many local Thai have been able to afford investing in the purchase of modern boxes. At present, new beekeeping management techniques meant beekeepers now have modern beekeeping knowledge and skills. However, European beekeeping techniques were applied to *A. cerana* without much success since their low honey production could not balance the cost of modern beekeeping costs and their frequent absconding into nature. The flexibility of European honey bees to adapt to a different environment has made them one of the most popular species in the beekeeping industry in Southeast Asia. Food and Agriculture Organization (FAO) data shows that the global population of European honey bee hives has increased by about 45% during the last 50 years due to economic globalization (Aizen and Harder 2009). As such, European honey bees are now regarded as the most widely distributed alien pollinator in the world (Kearns et al. 1998).

Fig. 8.5 European beekeeping in Chiang Mai, Thailand. Photo taken by the author

Even though *A. cerana* bees produce less honey than the European honey bee it is important to note that they are more resistant to diseases and pests (Chantawannakul et al. 2016), adapt well in the mountainous areas, and require lower levels of investment. Nowadays, about 6000 colonies are present in Thailand (Oldroyd and Wongsiri 2006). In addition, cavity nesting Asian honey bees have been reported to be more efficient pollinators of various fruit and vegetable crops than their European counterparts (Hepbrun and Radloff 2011; Rahman and Rahman 2000). Giant honey bees also have a wider foraging range than European honey bees and this can assist in the pollination of the forest ecosystem. In lowland forests, abundant with trees of the family Dipterocarpaceae, *A. dorsata*, the giant honey bee, is the main pollinator for at least 15 species of trees in Lambir (Momose et al. 1998) and the upper strata of forests in peninsular Malaysia (Appanah 1985). Dwarf Asian honey bees are also excellent as orchard and filed crop pollinators such as longan and mango in Thailand (Wongsiri et al. 1996). The roles of native honey bees in modern agro-ecosystems are also confirmed by the study showing that native bees can provide sufficient pollination for watermelon crops in the absence of domesticated honey bees (Ruttner 1988; Winfree et al. 2007). At present, bee products are honey, brood, propolis, bee wax, royal jelly, bee pollen, and bee venom and they are widely used in commercial products in the region (Oldroyd and Wongsiri 2006). Nevertheless, honey is still regarded more as a medicine than a food and bee products that are consumed by local people vary greatly across the region depending on local cultures and the sphere of religious influence. Royal jelly, pollen, bee venom, and propolis are known to be consumed as part of food supplements and medicinal alternatives.

Environmental Changes: Honey Bee Adaptation

Pests Parasites and Pathogens

As previously mentioned, honeybees are the most efficient of all the bees as pollinators of crops and natural flora (Hepbrun and Radloff 2011), however, in recent years, the decline of the honey bee population has become a great concern (Cox-Foster et al. 2007). Colony collapse disorder (or CCD) or bee losses are reported in many regions across North America and Europe (Genersch et al. 2010; van Engelsdorp et al. 2010), however, there are few reports of managed and non-managed colonies lost in Southeast Asia (Chantawananakul et al. 2016). The main factors that affect the honey bee populations are pesticide application in agricultural fields and changes of plant biodiversity (Oldroyd and Wongsiri 2006). Competition between honey bee species and inter-species transfer of pathogen and parasites have led to these losses (Yang 2005; Stout and Morales 2009; He and Lui 2011; Li et al. 2012).

European honey bees have encountered diseases and parasites which have co-evolved with wild native honey bees in Asia (Chantawannakul et al. 2016). The most serious and widespread parasites today are the *Varroa* mite and the *Nosema*, acclaimed to originate from *A. cerana*. They have been linked with colony loss in many parts of the world (Higes et al. 2008). In some parts of Southeast Asia, *Tropilaelaps* mites, originally parasitized in giant honey bees, are now probably the most destructive pests and prevalent in European beekeeping throughout the region (Chantawannakul et al. 2016). The number of infesting *Tropilaelaps* mites are often higher than *Varroa* mites in Thailand (Burgett et al. 1983; Buawangpong et al. 2015). Bee viruses are also reported to be jumping from native Asian honey bee species to the European honey bee when they are sharing the same habitat as with, for example, the Kashmir bee virus (KBV), first discovered in *A. cerana* (Bailey and Woods 1977). Likewise, the European honey bee also acts as a pest and pathogen reservoir that can later be transmitted back to native Asian honey bees (Chantawannakul et al. 2016), thereby promoting the presence of bee pests and parasites in the region.

Modern Society and Complications

Forests are an important source of livelihood for millions of people and play critical roles in regulating the world's climate, especially in Southeast Asia (Miura et al. 2015; Romijin et al. 2015). Honey bees depend on plant nectars and pollens for their carbohydrates, proteins, and fat diets. Plant resin is also collected by cavity nesting honey bees for their nest structure, propolis. This plays an important role in the defense of colonies due to their antimicrobial properties (Sanpa et al. 2015). Human agricultural activities and urban expansion have led to rapid deforestation, which is depriving honey bees of their food sources. However, few studies have shown the positive effects of urbanization on cavity nesting honey bees within urban areas (Winfree et al. 2007).

Deforestation is presently considered to be a major threat to loss biodiversity and habitat fragmentation (FAO 2010). Southeast Asia's mountainous mainland region stretches across Cambodia, Lao PDR, Myanmar, Thailand, Vietnam, and China and represents the largest tropical forest in the mainland Southeast Asia (Rerkasem et al. 2009). At present, wood removal is highest in many Southeast Asian countries and is on the increase. FAO data indicate that Indonesia and Myanmar were ranked in the top ten countries in the world that emitted CO_2 from land use during the years 1990–2012 (FAOSTAT 2015). With rapid deforestation and with permanent cropping, land use in the region is rapidly transforming landscapes. Thus, forest areas or natural habitat for wild bees have been greatly reduced. The expansion of cash crops (e.g. rubber, cabbage, corn, and maize) are also greatly increasing and acting as a catalyst for change (Ahrends et al. 2015). Farmers in the region have been given seeds and fertilizers on credit and the traders will come to purchase their harvest (Vongvisouk et al. 2016). Market networks have been responsible for introducing modern technology to the region. However, the downturn of this operation is that villagers tend to invade forest areas and practice monoculture-dominated agriculture, greatly reducing plant diversity. This results in the decline of food sources and habitats for wild bees. Each honey bee species may be affected by deforestation to different degrees. Deforestation greatly affects the cavity Asian honey bee, *A. cerana*, as old-growth forests provide cavity containing trees for nesting sites (Oldroyd and Wongsiri 2006). *A. koschevnikovi* is restricted to intact rainforest as in Borneo (Otis 1996). However, *A. florea* which does not need cavity for nesting, seems to adapt well in disturbed habitats (Oldroyd and Wongsiri 2006). Regarding previous observations, *A. dorsata* has been reported to be in decline due to the clearing of Asian Dipterocarp forests (Oldroyd and Wongsiri 2006). Likewise, *A. laboriosa* population is declining in Nepal because of deforestation in the middle-hill zone between 1000–2500 m (Underwood 1992; Thapa 2001; Valli and Summers 1998). In addition, the negative impacts of modern agricultural interventions, e.g., the use of chemical fertilizers, pesticides, and environmental pollution have been reported in many countries including Southeast Asia and are a major source of the decline of honey bee populations (Aizen and Fiensinger 1994; Partap and Partap 1997, 2002; Ricketts et al. 2008).

Invasive plant species may also have additional direct or indirect impacts on feral honey bees. Pollen varies in protein content amongst plant species and this affects the pollen foraging behavior of honey bees. If they are a higher quantity and quality of invasive species than native plants, this could act as a food source. In Northern Thailand, a sensitive plant (*Mimosa pudica*) is one of the major pollen sources for honey bees. On the other hand, if pollens and nectars of the invasive plant species are not suitable for honey bees they can also have a negative impact on honey bee survival (Stout and Morales 2009).

Climate change is also another factor that might impact honey bee diversity in the region (Brown and Paxton 2009). Seasonal shift, rainfall, drought, distribution/extinction of flower species, and other environmental factors are impacting on nectar and pollen flows, the main food sources for honey bees (Thuiller et al. 2005; Le Conte and Navajas 2008). Honey bee species (*A. laboriosa, A. nigrocincta, and A. nuluensis*) that have narrow range of habitat will be greatly affected by climate change as their access to floral sources in their habitats becomes restricted (Oldroyd and Wongsiri 2006). Meteorological modeling forecasts that the average global temperature will increase 1–6 °C over the next hundred years. If this occurs, it will affect the weather patterns including the frequency of extreme weather and impact upon pollinators such as honey bees (IPCC 2001). Therefore, the key risks of the region are warming and drying trends, extreme temperatures, extreme precipitation, and cyclones. Additionally, the expansion of cities and new job opportunities in urban areas has seen many people move from rural areas and leave their agricultural fields impacting upon the maintenance of bee colonies.

Although there are individual pockets of successful European beekeeping operations in the region, the industry is still impeded in some countries (e.g., Laos, Cambodia, and Myanmar) by several factors. These include a lack of infrastructure, a lack of large-scale agriculture, and necessary beekeeping management knowledge. European beekeeping usually requires capital investments and external inputs which are not easy for local communities to initiate. In particular, management costs and apicultural practices are somewhat beyond the economic means of most local people (fixed cost for beehive equipment 1600 THB (US $48)/hive and the operational cost is 2000 THB/hive/year (US $57) (US $1 = 35 THB)). In addition, beekeeping operations are sometimes not economically viable (Chantawannakul et al. 2004). Surveys show that the most viable beekeeping operations in Chiang Mai, Thailand are medium-sized ones (101–1000 hives), which account for about 66% of the total number of beekeepers. Colony operation costs include local bee pest and parasite controls, sugar that is used to feed colony during dearth periods, land rental costs, and transport expenses for trucks and other heavy equipment due to the migratory nature of the business. Small-scale beekeeping operations remain as sideline enterprises, whereas medium- and large-scale operations can develop into full-time occupations. The pollination service business does not exist as it does in other western countries such as the US. Nevertheless, other factors such as the input of capital must be taken into consideration. Beekeeping management techniques, marketing practices, and seasonal factors such as climate, nectar flow, etc. are important variables as they influence the investment and decision-making of local beekeepers not just in Northern Thailand, but also in other parts of Southeast Asia.

Some local beekeepers cannot obtain high yields of honey. Factors that contribute to low honey yields are a lack of beekeeping knowledge, diseases, and pests, as well as agricultural pollutants. In some parts of Southeast Asia, domestic supply does not meet national demand. As such, shortfalls are met by imports. According to published reports, exports of packaged honey from Western Australia to Malaysia and Singapore in particular, increased over a five-year period until the Asian economic downturn (Chantawannakul et al. 2004).

Future Prospects

Most world regions including Southeast Asia, are experiencing the risks and opportunities of economic globalization. Human agro-economic activities not only affect the global bee population but in some cases, also can disrupt the evolutionary process of honey bee species. Habitat change and the loss and fragmentation associated with agriculture intensification are fundamental risks for native honey bees (Kremen et al. 2002; Murray et al. 2009). Therefore, the conservation of honey bee diversity and their ecosystems is the greatest challenge that we have ever encountered. The "pollinator crisis" is not far-fetched. Honey bee losses, in turn, will affect both wildflowers and cultivated crops, which are dependent on pollination service by honey bees (Potts et al. 2010). The increase of European honey bee populations in the region for commercial purposes might compensate pollination service, provided by feral honeybees, however, European honey bee populations will not cover all. Introduced honey bee species, in this case, the European honey bee has never been thoroughly assessed in Southeast Asia, in terms of food and nest site competition and this requires urgent research attention. Environmental costs warrant recognition and consideration during the development of agriculture and conservation policies, and also effective measures that can put policies into practice. As mentioned above, the conservation of honey bee species in Southeast Asia not only depends on biological factors but also social, political, and economic ones. New technologies have introduced new regimes of extraction and commodification that directly impact on the region's key pollinators. The beekeeping industry in Southeast Asian countries requires a holistic approach including input from scientists, policy makers, beekeepers, and business partners. As such, a constructive and innovative *modus operandi* is essential to promote the economic vitality of the beekeeping industry and also maintaining the balance of wild bees to warrant sufficient pollinators in the forest ecosystem and guarantee human food sources as human societies and their activities intensively impact upon the region's ecosystems and services.

Acknowledgements The author would like to thank the Chiang Mai University fund and Thailand research fund (RSA 6080028) for supporting this research.

References

Ahrends, A., Hollingsworth, P. M., Ziegler, A. D., Fox, J. M., Chen, H., Su, Y., & Xu, J. (2015) Current trends of rubber plantation expansion may threaten biodiversity and livelihoods. *Global Environmental Change, 34,* 48–58.

Aizen, M. A., & Harder, L. D. (2009). The global stock of domesticated honey bees is growingslower than agricultural demand for pollination. *Current Biology, 19,* 915–918.

Aizen, M. A., & Feinsinger, P. (1994). Ecological and genetic effects on demographic processes: Pollination, clonality, and seed production in *Dithyrea maritima. Biological Conservation, 116,* 27–34.

Appanah, S. (1985). General flowering in the climax rain forests of South East Asia. *Journal of Tropical Ecology, 1,* 225–240.

Bailey, L., & Woods, R. D. (1977). Two more small RNA viruses from honey bees and further observations on sacbrood and acute bee-paralysis viruses. *Journal of General Virology, 37*(1), 175–182.

Bankova, V., de Castro, S., & Marcucci, M. (2000). Propolis: Recent advances in chemistry and plant origin. *Apidologie, 31,* 3–15.

Brown, M. J. F., & Paxton, R. J. (2009). The conservation of bees: A global perspective. *Apidologie, 40*(3), 410–416.

Buawangpong, N., Saraithong, P., Khongphinitbunjong, K., Chantawannakul, P., & Burgett, M. (2014). The comb structure of *Apis dorsata* F. (Hymenoptera: Apidae): 3-dimensional architecture and resource partitioning. *Chiang Mai Journal of Science, 41,* 1077–1083.

Buawangpong, N., de Guzman, L. I., Frake, A. M., Khongphinitbunjong, K., Burgett, M., & Chantawannakul, P. (2015). *Tropilaelaps mercedesae* and *Varroa destructor*: prevalence and reproduction in concurrently infested *Apis mellifera* colonies. *Apidologie, 46*(6), 779–786.

Buchmann, S. L., & Nabhan, G. P. (1996). *The forgotten pollinators*. Washington D.C.: Island Press.

Burgett, D. M., Akratanakul, P., & Morse, R. (1983). *Tropilaelaps clareae*: A parasite of honey bees in South East Asia. *Bee world, 64,* 25–28.

Chantawannakul, P., Petersen, S., & Wongsiri, S. (2004). Conservation of honey bee species in South East Asia: *Apis mellifera* or native bees? *Biodiversity, 5*(2), 25–28

Chantawannakul, P., de Guzman, L. I., Li, J., & Williams, G. R. (2016). Pests, pathogens, and parasites of honey bees in Asia. *Apidologie, 47,* 321–324. https://doi.org/10.1007/s13592-015-0407-5.

Chen, P. P., Wongsiri, S., Jamyanya, T., Rinderer, T. E., Vongsamanode, S., Matsuka, M., Sylvester, H. A., & Oldroyd, B. P. (1998). Honey bees and other edible insects used as human food in Thailand. *American Entomologist, 44,* 24–29.

Crane, E. (1999). *The world history of beekeeping and honey hunting*. London: Gerald Duckworth & Co. Ltd.

Cockerell, T. D. A. (1907). A fossil honey-bee. *Entomologist, 40,* 227–229.

Cockerell, T. D. A. (1909). Some European fossil bees. *Entomologist, 42,* 313–317.

Cox-Foster, D. L., Conlan, S., Holmes, E. C., Palacios, G., Evans, J. D., Moran, N. A., et al. (2007). A metagenomic survey of microbes in honey bee colony collapse disorder. *Science, 318*(5848), 283–287.

Crane, E. (2003). Making a bee-line international bee research association. *International Bee Research Association*, Cardiff, United Kingdom.

Engel, M. S. (1998). Fossil honey bees and evolution in the genus *Apis* (Hymenoptera: Apidae). *Apidologie, 29,* 265–281.

Fabricius, J. C. (1787). *Mantissa insectorum systmeseorum species nuper detectas adiectis characteribus genericus, differentitis specificis, emendationibus observationibus* (Vol. 1). [Supplemental list of insects setting out recently discovered species together with the generic characteristics, specific distinguishing features, emendations and remarks]. Hafniae, Proft, Copenhagen, Denmark.

Fabricius, J. C. (1793). *Mantissa insectorum systemseorum species nuper detectas adiectis characteribus genericus, differentitis specificis, emendationibus, observationibus* [Supplemental list of insects setting out recently discovered species together with the generic characteristics, specific distinguishing features, emendations and remarks]. Hafniae, Proft, Copenhagen, Denmark.

FAOSTAT data. (2015). Retrived November, 2015. http://faostat3.fao.org/browse/G2/*/E.

Food and Agriculture Organization of the United Nations (FAO) (2010) FAOSTAT. (http://faostat.fao.org).

Genersch, E., Evans, J. D., & Fries, I. (2010). Honey bee disease overview. *Journal of Invertebrate Pathology, 103*(SUPPL. 1), S2–S4.

Hadisoesilo, S. (2000). Tingku, a traditional management technique for *Apis dorsata Binghami* in central Sulawesi, Indonesia. Seventh IBRA conference and Fifth AAA conference, Chiang Mai, Thailand, 141.

He, X., & Liu, X. Y. (2011). Factor of *Apis ceranae* decline in China. *Apiculture of China, 62*(5), 21–23.

Hepbrun, R., & Radloff, S. E. (2011). *Honeybees of Asia* (p. 669). Berlin: Springer.

Higes, M., Martín-Hernández, R., Botías, C., Garrido-Bailón, E., González-Porto, A. V., Barrios, L., et al. (2008). How natural infection by *Nosema ceranae* causes honeybee colony collapse. *Environmental Microbiology, 10,* 2659–2669.

IPPC. 2001. Climate change: the scientific basis. In *Technical summary from working group I.* Geneva: Intergovernmental Panel on Climate Change.

Kearns, C. A., Inouye, D. W., & Waser, N. M. (1998). Endangered mutualisms: the conservation of plant-pollinator interactions. *Annual Review of Ecology, Evolution, and Systematics, 29,* 83–112.

Klein A. M., Vaissière, B., Cane, J. H., Steffan-Dewenter, I., Cunningham, S. A., Kremer, C., & Tscharntcke, T. (2007). Importance of pollinators in changing landscapes for world crops. *Proceeding of the Royal Society of London B, 274,* 303–313

Kobayashi, S. (2013). A study of Buddhist places of worship in rural Cambodia: With a special focus on their differences and formative processes. *Southeast Asian Studies, 51*(1), 34–69.

Kremen, C., Williams, N. M., & Thorp, R. W. (2002). Crop pollination form native bees at risk from agricultural intensification. *Proceedings of the National Academy of Sciences of the United States of America, 99,* 16812–16816.

Le Conte, Y., & Navajas, M. (2008). Climate change: Impact on honey bee populations and diseases. *Revue Scientifique et Technique, 27*(2), 485–510.

Li, J. Qin, H., Wu, J., Sadd, B.M., Wang, X., Evans J.D., & Peng W. (2012). The prevalence of parasites and pathogens in Asian honey bees *Apis cerana* in China. *PLOS ONE* 7(11), e47955. 1–12

Linnaeus, C. (1758). *Systema naturae per regna tria naturae, secundum classes, ordines, genera, species, cum characteribus, differentiis, synonymis, locis* [System of nature through the three kingdoms of nature, according to classes, orders, genera and species, with characters, differences, synonyms, places]. Stockholm, Sweden.

Lo, N., Gloag, R. S., Anderson, D. L., & Oldroyd, B. P. (2010). A molecular phylogeny of the genus *Apis* suggests that the giant honey bee of the Philippines, *A. breviligula* Maa, and the plains honey bee of southern India, *A. indica* Fabricius, are valid species. *Systematic Entomology, 35*(2), 226–233

Miura, S., Amacher, M., Hofer, T., San Miguel, S., Ernawati, & Thackway, R. (2015). Protective functions and ecosystem services of global forests in the past quarter-century. *Forest Ecology and Management, 352*, 109–123.

Momose, K. R. R., Ishii, S., Sakai, S., & Inoue, T. (1998). Reproductive intervals and pollinators of tropical plants. *Proceeding of the Royal Society of London B, 265,* 2333–2339.

Murray, T. E., Kuhlmann, M., & Potts, S. G. (2009). Conservation ecology of bees: Populations, species and communities. *Apidologie, 40,* 211–236.

Oldroyd, B., & Wongsiri, S. (2006). *Asian honey bees: Biology, conservation, and human Interactions*. Cambridge: Harvard University Press.

Otis, G. W. (1996). Distribution of recently recognized species of honey bees (Hymenoptera: Apidae, *Apis*) in Asia. *Journal of the Kansas Entomological Society, 69,* 311–333.

Partap, U., & Partap, T. (1997). Managed crop pollination: The missing dimension of mountain agricultural productivity. Mountain farming systems. Discussion Paper Series No. MFS 97/1. Kathmandu, Nepal International Centre for Integrated Mountain Development (ICIMOD).

Partap, U., & Partap. T. (2002). Warning signals from apple valleys of the HKH region: Pollination problems and farmers' management efforts. Kathmandu, Nepal International Centre for Integrated Mountain Development (ICIMOD).

Pattamayutanon, P., Angeli, S., Thakeow, P., Abraham, J., Disayathanoowat, T., & Chantawannakul, P. (2015). Biomedical activity and related volatile compounds of Thai honeys from three different honeybee species. *Journal of Food Science, 80*(10), M2228–M2240.

Potts, S. G., Biesmeijer, J. C., Kremen, C., Neumann, P., Schweiger, O., & Kunin, W. E. (2010). Global pollinator declines: trends, impacts and drivers. *Cell Press, 25*(6), 345–353.

Rahman, A., & Rahman, S. (2000). Effect of honey bee (*Apis cerana indica*) pollination on seedset and yield of buckwheat (Fagopyrum esculentum). *The Indian Journal of Agricultural Sciences, 70,* 168–169.

Rerkasem, K., Lawrence, D., Padoch, C., Schmidt-Vogt, D., Zeigler, A. D., & Bruun, T. B. (2009). Consequences of swidden transitions for crop and fallow biodiversity in Southeast Asia. *Human Ecology, 37*(3), 347–360.

Ricketts, T. H., Regetz, J., Steffan-Dewenter, I., Cunningham, S. A., Kremen, C., Bogdanski, A., et al. (2008). Landscape effects on crop pollination services: are there general patterns? *Ecology Letters, 11*(5), 499–515.

Romijin, E., Lantican, C. B., Herold, M., Lindquist, E., Ochieng, R., Wijaya, A., et al. (2015). Assessing change in national forest monitoring capacities of 99 tropical countries. *Forest Ecology Management, 352,* 109–123.

Rosenzweig, M. L. (1995). *Species diversity in space and time* (p. 436). Cambridge: Cambridge University Press.

Roussy, L. (1960). Insect et abeilles de l' amber de Sicile [The amber insects and bees of Sicily]. *Gazette Apic, 635,* 5–8.

Ruttner, F. (1988). *Biogeography and taxonomy of honeybees*. Berlin: Springer.

Sanpa, S., & Chantawannakul, P. (2009). Survey of six bee viruses using RT-PCR in Northern Thailand. *Journal of Invertebrate Pathology, 100,* 116–119.

Sanpa, S., Popova, M., Bankova, V., Tunkasiri, T., Eitssayeam, S., & Chantawannakul, P. (2015). Antibacterial compounds from propolis of *Tetragonula laeviceps* and Tetrigona melanoleuca (Hymenoptera: Apidae) from Thailand. *PLoS ONE, 10*(5), e0126886.

Sanpa, S., Popova, M., Tunkasiri, T., Eitssayeam, S., Bankova, V., & Chantawannakul, P. (2016). Chemical profiles and antimicrobial activities of Thai propolis collected from *Apis mellifera. Chiang Mai Journal of Science, 44*(2), 438–448.

Saraithong, P., Burgett, M., Khongphinitbunjong, K., & Chantawannakul, P. (2012). *Apis dorsata* F.: Diurnal foraging patterns of worker bees in northern Thailand. *Journal of Apicultural Research, 51*(4), 362–364.

Smith, F. (1858). Catalogue of the hymenopterous insects collected at Sarawak, Borneo. Mount Ophir, Malacca, and at Singapore. *Proceedings* of the *Linnean Society* of *London Zoological, 2,* 42–130.

Smith, F. (1861). Description of new species of hymenopterous insects collected by Mr. A.R. Wallace at Celebs. *Proceedings of the Linnean Society, London, 5,* 57–93.

Smith, F. (1871). Description of some new insects collected by Dr. Anderson during the Expedition to Yunnan. *Proceeding of the Zoological Society of London, 1871,* 169–171.

Spellerberg, I. F., & Sawyer, J. W. D. (1999). *An introduction to applied biogeography* (p. 243). Cambridge: Cambridge University Press.

Stauffer, P. H. (1979). A fossilized honeybee comb from Late Cenozoic cave deposits at Batu Caves, Malay Peninsula. *Journal of Paleontology, 53,* 61416–611421.

Stout, J. C., & Morales, C. L. (2009). Ecological impacts of invasive alien species on bees. *Apidologie, 40,* 388–409.

Thapa, R. (2001). The Himalayan giant honey bee and its roles in ecotourism development in Nepal. *Bee world, 82,* 139–141.

Thuiller, W., Lavorel, S., Araújo, M. B., Sykes, M. T., & Prentice, I. C. (2005). Climate change threats to plant diversity in Europe. *Proceedings of the National Academy of Sciences of the United States of America, 102,* 8245–8250.

Théobald, N. (1937). *Les insects fossiles des terrains oligocenes de France* [Fossil Insects of Oglicene France]. France: George Thomas, Nancy.

Tingsek, S., Koeniger, G., & Koeniger, N. (1996). Description of a new cavity nesting species of *Apis* (*Apis nuluensis* n. sp.) from Sabah, Borneo, with notes on its occurrence and reproductive biology. *Senckenbergiana Biologica, 76,* 115–119.

Tingsek, S., Mardan, M., Rinderer, T. E., & Koeniger, G. (1988). Rediscovery of Apis vechti (Maa, 1953): the Saban honey bee. *Apidologie, 19,* 97–102.

Underwood, B. A. (1992). Impact of human activities on the Himalayan honey bee, *Apis*. In L. R. Verman (Ed.), *Honey bees in mountain agriculture* (pp. 51–58). Boulder: Westview Press.

Valli, E., & Summers, D. (1998). *Honey hunters of Nepal*. Abrams, New York: Harry N.

van Engelsdorp, D., Speybroeck, N., Evans, J. D., Nguyen, B. K., Mullin, C., Frazier, M., F Frazier, J., Cox-Foster, D., Chen, Y., Tarpy, D. R., Haubruge, E., Pettis, J. S., & Saegerman, C. (2010). Weighing risk factors associated with bee colony collapse disorder by classification and regression tree analysis. *Journal of Economic Entomology, 103* (5), 1517–1523.

van Huis, A., Van Itterbeeck, J., Klunder, H., Mertens, E., Halloran, A., Muir, G., & Vantomme, P. (2013). *Edible insects—Future prospects for food and feed security*. FAO Forestry Paper 171.

Vongvisouk, T., Broegaard, R. B., Mertz, O., & Thongmanivong, S. (2016). Rush for cash crops and forest protection: Neither land sparing nor land sharing. *Land use policy, 55,* 182–192.

Wanjai, C., Sringarm, K., Santasup, C., Pak-Uthai, S., & Chantawannakul, P. (2012). P Physicochemical and microbiological properties of longan, bitter bush, sunflower, and litchi honeys produced by *Apis mellifera* in Northern Thailand. *Journal of Apicultural,* 51(1), 36–44.

Waring, C., & Jump, D. R. (2004). Rafter beekeeping in Cambodia with Apis dorsata. *Bee World, 85*(1), 14–18. https://doi.org/10.1080/0005772X.2004.11099607.
Winfree, R., Williams, N. M., Dushoff, J., & Kremen, C. (2007). Native bees provide insurance against ongoing honey bee losses. *Ecology Letters, 10*(11), 1105–1113.
Wongsiri, S., Chen, P., & Thapa, R. (1995). Other uses of honey bee-products in Thailand. *British Bee Journal, 123,* 144–149.
Wongsiri, S., Thapa, R., Oldroyd, B. P., & Burgett, D. M. (1996). A magic bee tree, home to *Apis dorsata. American Bee Journal, 136,* 95–98.
Wongsiri, S., Tangkanasing, P., & Sylvester, H. A. (1987). Mites, pests and beekeeping with *Apis cerana* and *Apis melifera* in Thailand. *American Bee Journal, 127,* 500–503.
Yang, G.-H. (2005). Harm of introducing the western honeybee *Apis mellifera* L. to the Chinese honeybee *Apis cerana* F. and its ecological impact. *Acta Entomologica Sinica, 48,* 401–406
Yap, P., Bakar, M. F. A., Lim, H., & Carrier, D. (2015). Antibacterial activity of polyphenol-rice extact of selected wild honey collected in Sabah, Malaysia. *Journal Apicultural Research, 54* (3), 163–172.
Zeuner, F. E., & Manning, F. J. (1976). A monograph on fossil bees (Hymenoptera: Apoidea). *Bulletin of the British Museum (Natural History), 27,* 151–268.
Zhang, J. F. (1990). New fossil species of Apoidea (Insecta: Hymenoptera). *Acta Zootax Sinica, 15,* 83–91.

Part III
The Politics of Energy and Resources-use

Chapter 9
The Rise of Organic Agriculture in the Philippines and its Development

Bao Maohong

Abstract This chapter examines organic agriculture as a definite alternative to industrial forms of agriculture in the Philippines. It argues that its mainstreaming depends on the complex and difficult reconstruction of socioeconomic relations and the political system of the country. In recent years, organic agriculture in the Philippines has become an emergent market integrating into the national economy. In 1980s, the green revolution was proclaimed a failure due to its economic unsustainability and negative impact on the environment. From this and through a combination of the rise of the international organic agriculture movement and the reinvention of traditional Philippine agriculture, organic agriculture has come to play a role as an alternative form of agricultural development. This chapter argues that the conscious development of organic agriculture is a response to the failures of the present agricultural structure and offers a potentially economically viable alternative to rehabilitate degraded arable land.

Keyword Philippines · Organic agriculture · Green revolution
Industrial agriculture · International federation of organic agriculture movements (*IFOAM*) · National organic agriculture program (*NOAP*)

Introduction

Agriculture is one of the main sectors of the Philippine national economy and contributes 17% to the Philippine gross domestic product, employs 33% of the country's labor force, and accounts for an estimated 66% of the country's poorest.[1] It has been commercialized and modernized as early as 1780s and since, independence in 1946, various nationalist governments have aimed at promoting

[1]See: National Organic Agriculture Board (NOAB) (2012, p. 10).

B. Maohong (✉)
History Department, Peking University, Yiheyuan Road 5, Beijing 100871, China
e-mail: baomh616@pku.edu.cn

M. Lopez and J. Suryomenggolo (eds.), *Environmental Resources Use and Challenges in Contemporary Southeast Asia*, Asia in Transition 7,
https://doi.org/10.1007/978-981-10-8881-0_9

industrialization underpinned by agriculture, although import substitution and export orientation strategies have been implemented. In this context, Philippine agriculture has undergone different periods of industrialization. However, export-oriented agriculture did not satisfy domestic demands for food, but also resulted in serious environmental problems. As such, organic agriculture emerged in response to these challenges at the right moment and is leading to the transformation of the country's agricultural sector.

According to International Federation of Organic Agriculture Movements (IFOAM), organic agriculture is a production system that sustains the health of soils, ecosystems, and people. It relies on ecological processes, biodiversity and cycles adapted to local conditions, rather than the use of inputs that have adverse effects. Organic agriculture combines tradition, innovation, and science to benefit the shared environment and promote fair relationships and a good quality of life for all involved.[2]

According to Republic Act No. 10068 (Organic Agriculture Act), organic agriculture includes all agricultural systems that promote the ecologically sound, socially acceptable, economically viable, and technically feasible production of food and fibers. Organic agriculture dramatically reduces external inputs by refraining from the use of chemical fertilizers, pesticides, and pharmaceuticals. It also covers areas such as, but not limited to, soil fertility management, varietal breeding and selection under chemical and pesticide-free conditions, and the use of biotechnology and other cultural practices that are consistent with the principles and policies of this Act. Furthermore, it enhances productivity without destroying the soil and harming farmers, consumers, and the environment as defined by IFOAM. The biotechnology herein referred to shall not include genetically modified organisms.[3] What we see here is that the Philippine definition of organic agriculture is, in fact, more concrete and practical than that offered by IFOAM.

This chapter addresses the background of the emergence of organic agriculture in the Philippines, its dynamics, and development models and states in three parts. First, it starts with a brief overview on the history of agriculture in the Philippines from the Spanish colonial period to the independence period and up to 1980s to contextualize a dominant agricultural policy of the time that did not include organic farming practices. Second, it describes how organic agriculture arose in the Philippines and became a part of government policy in the 1990s. Third, through fieldwork, it focuses on the latest developments in Philippine organic agriculture in the early twenty-first century. Last, it concludes with a discussion on the future challenges that organic agriculture faces in the Philippines.

---

[2]For the definition of organic agriculture, please see: http://www.ifoam.bio/en/organic-landmarks/definition-organic-agriculture (Accessed February 2, 2015). This definition is formulated in the IFOAM general assembly in Italy, in 2008.

[3]See: Republic Act No. 10068, known as Organic Agriculture Act of 2010. http://www.da.gov.ph/index.php/laws-issuances (Accessed February 2, 2015).

The Emergence of the Philippines' Organic Agriculture

It goes without saying that agriculture in the Philippines remained highly underdeveloped creating serious food shortages during the colonial period (1521–1898) (Veneracion 2000, p. 206). During the Spanish colonial period, colonizers tried to transplant their own agriculture and pasture model into the Philippines archipelago (Veneracion 2000, pp. 49–84). However, this attempt evidently failed, given the fact that climate and geographical conditions in the Philippines were radically different from those in Europe (Veneracion 2000, pp. 39–84). Subsequently, the colonial state focused on the Manila galleon trade, ignoring agricultural development which resulted in economic stagnation in the Philippines. In the late eighteenth century, however, the Spanish colonial state started to reform its colonial policy and open up agriculture to the world market (Veneracion 2000, pp. 111–152). With investment, mainly from the UK, Germany, and the USA, Philippine agriculture transformed into and became a part of the global cash cropping plantation system albeit one adapted to the local natural environment (Corpuz 1997, pp. 87–138). This resulted in the integration of the Philippines' agricultural system into the capitalist world-economy. After independence in 1946, however, an unbalanced agricultural production system did not significantly nor efficiently change even as the national government supported a mode of industrialization based on earnings from agricultural product exports. Concurrently, this increased added value through agricultural products processing. Nonetheless, the essential goal was to reach self-sufficiency in the production of grain.

At the time, there were two parallel models practiced by the Philippine government: a denotative (extensive) development model (in the late Spanish colonial period through to postindependence) and a connotative (intensive) development model (practiced by the ministry of agriculture and IRRI). The first was to expand cultivated areas available for food crops, whereas the second launched the green revolution through which yields per unit in the major grain producing areas of the country would be improved by increasing capital and technology investment.

Expanding cultivated areas of food crops was the most common way of boosting the country's agricultural production. With an increase in population growth[4] after independence, cultivated areas of food crops had to be expanded to respond to urgent demands for food and employment. Due to overpopulation on coastal farmlands and alluvial plains that were traditional cultivated areas, landless populations were forced to move to upland areas and turn forest areas into farmland for subsistence (AAS 1983, p. 140) Following commercial logging corporations, lowland landless peasants migrated and cultivated the remnants of forest through sedentary cultivation technology. Consequently, the upland *Kaingineros* (swidden farmers) were forced to move into steeper areas and shorten their fallow periods in swidden cultivation. In order to improve production, cultivated areas of food and

[4]From 1948 to 1990, the population grew from 19.254 million to 62 million, with an average yearly growth rate of 2.58%. See Mercedes B. Concepcion, The Philippine population problem: Myth and reality, *Philippine Studies*, Vol.42, No.2, 1994, P.139.

main cash crops (such as corn, sugarcane, banana, pineapple, mango, coffee, etc.) increased from 5 million hectares in 1962 to 10 million in 1985 (Boyce 1993, p. 64 and 70). As a result of these land changes, the Philippines now experiences a large number of natural and man-made disasters. Large concentrations of annual precipitation have compounded manmade transformations of the environment. Soil erosion, mountain landslides, and urban flooding have forced peasant populations to further encroach and cultivate other forest land and steeper uplands. This development model has resulted in a vicious cycle of poverty, development, and environmental deterioration. This form of development has been unsustainable for livelihood and has not led to substantial economic improvements for peasants' livelihood, whilst eroding the potential of future upland development.

The second, implemented in the 1960s, was the connotative (intensive) development model and focused on improving yields per unit. Combined with official assistance strategy from postwar USA, the internal claims for improving agricultural production produced the green revolution in Mexico, India, and the Philippines, and other developing countries in the region. As the most important grain crop, rice planting was promoted by the Philippines' government alongside the development and introduction of high-yielding varieties and agricultural technology. Supported by the Rockefeller foundation and Ford foundation, in 1962, the International Rice Research Institute (IRRI) was set up in the Philippines, and aimed at breeding a new variety that combined the different good qualities of temperate rice and tropical rice (norin no. 1). In 1966, scientists developed and bred IR 8, a semi-dwarf variety that was lodge-resistant, of stout-stature, responsive to fertilizer, and high-yielding. Boyce (1993) notes that "(b)y the 1970–1971 crop year, half of the country's rice acreage was planted with the new varieties, and this rose to 75% by 1979–1980" (Boyce 1993, p. 62). Rice yields rose at 3.4% per year. Between 1962–1964 and 1983–1985 rice yields doubled from 1.24 to 2.48 metric tons of *palay* (rice) per hectare (Boyce 1993, p. 63). The green revolution in the Philippines succeeded in obtaining its primary objective (rice self-sufficiency) allowing the Philippines to become a rice exporter country. However, the sharp yield increases from the green revolution were exhausted by 1982,[5] and fast rice yield growth declined.

Although the green revolution produced yield growth, it did not result in income growth for the rural poor and the benefits of the green revolution did not trickle down to the landless poor. The overuse of fertilizer and pesticides (fertilizer accounted for 20% of all input in rice production, pesticides 16%) (see Boyce 1993), and monoculture brought about serious environmental pollution and ecological damage. The intensive planting of high-yielding varieties weakened genetic diversity, thus without natural enemies, it was easier for pests to rapidly spread to susceptible plants. Once insect infestation and diseases broke out, rice yields decreased dramatically. The other problem associated with pesticide use was the

[5]See: World Bank, *Philippines: toward sustaining the economic recovery*, Report No. 7438-PH, January 30, 1989, p.viii.

emergence of resistant pests. This resulted in the invention and application of more toxic pesticides and the overuse of these directly undermined the health of agricultural laborers. This subsequently impacted upon the aquatic food chain, polluted by pesticides, harming Filipino's health (FEA 2012). In central Luzon, the overuse of pesticides was followed by a 27% increase in deaths among farmers from causes other than physical injury (Briones 2005, p. 71). According to the Organic Producers Trade Association (OPTA), consuming the high-yielding agri-produce or the so-called "green revolution crops" caused brain damage and impaired cognitive function (FEA 2012). Furthermore, OPTA recognized international studies discussing the negative impact of chemical-infused crops: those that result in cancer, hormone disruption, neurological disorders, and other life-threatening illnesses.[6] Dependence on chemical fertilizer resulted in soil compaction and a decrease of its self-regulating ability. Additionally, the price of fertilizer increased with the price of oil in the world energy market, because most nitrogenous fertilizer was made of fossil fuels. Between 1966–1974, the real price of rice rose, and thereafter dropped sharply, one that was not favorable for producers (Boyce 1993, p. 101). Under such a situation, rice production in the Philippines fell into crisis, especially in terms of its economic and ecological sustainability.

As a result, a structural problem emerged in the Philippines' agriculture, one that increased in seriousness with rapid population growth. One way to reassess and respond to this problem is to readjust the mode of agriculture, particularly, relations between farmers and the land after its frontiers are reorganized. This kind of agriculture should be a new type that harmonizes and ties agricultural production to land carrying capacity. In 1985, the report "The Miracle that Never Was" by the Agency for Community Education and Services was presented at the National Convention of Rice Farmers, in which sustainable agriculture that harmonizes production with an environment based on ecological principles was recognized as the only feasible way. Subsequently, in 1985, protests by rice farmers against IRRI broke out in some areas. In one sense, this symbolized the beginning of the Philippine organic agriculture movement (see Carating and Tejada 2012).

Philippine Organic Agriculture Dynamism and its Development Models

The breakthrough to the agricultural dilemma originated from two different but amalgamated forces: the spread of the international organic agriculture movement, and the reinvention of the traditional Philippine agriculture. The former introduced organic agriculture that went beyond modern industrial models, whereas the *zeitgeist* of the latter introduced an invigorated indigenous organic agriculture.[7]

[6]See: "Organic farming: the future of Philippine agriculture," http://www.fareasternagriculture.com/crops/agriculture/organic-farming-the-future-of-philippine-agriculture (Accessed August 25, 2015).

[7]Some estimates have suggested the local organic product market was worth $US 6.2 million in 2000 and between $US 20 ~ 30 million in 2011 (National Organic Agriculture Program (NOAP) 2012–2016, p. 14.)

The international organic agriculture movement introduced some benefits. First, there were two favorable advantages for the Philippines to develop organic agriculture: a large world market of organic products, and a successful experience of developing organic agriculture. The Department of Trade and Industry (DTI) was responsible for introducing the country's natural and processed organic products into the international organic produce market. In 1986, the Center for International Trade and Exposition Missions of Department of Trade and Industry participated in the world organic trade fair at Nuremberg, Germany, and displayed natural organic products, processed, and frozen organic products from the Philippines. Since then, it has captured a sizeable share of the overseas global organic market. Attracted by this fair, Philippine farmers and businessmen began to be actively involved in organic agriculture through discovering the potential and economic value of Philippine organic products in the world market (see Carating and Tejada 2012). Thereafter, the Center participated in a number of fairs held in Japan and the USA to expand the market of the Philippines' organic products and gained a larger share of the world market.

Through joining in and organizing the Asia Scientific Conference of International Federation of Organic Agriculture Movement (IFOAM), local non-government organizations on organic farming in the Philippines have absorbed the successful experiences of international organic agriculture and made it accessible for the farmers.[8] At the first IFOAM Asian Conference in Saitama in 1993, delegates from the Philippines learned more from the ten principles of "TEIKEI" that the Japan Organic Agriculture Association showed and was campaigning for.[9] At the 1995 IFOAM-Asia Conference held in Seoul, Republic of Korea, not only did the Philippines' delegates understand the activities, achievements, and experiences of South Korean organic agriculture, but also presented a Philippine Standard for Organic Production and Processing based on IFOAM standards that underpinned the establishment of certification system of national organic agriculture. Furthermore, the Philippines' delegates, who attended the third IFOAM-Asia Conference in Banglore in 1997 understood the Indian recovery of indigenous organic agriculture. This improved their confidence to empower Philippine traditional agriculture. In 1999, the Philippine hosted the fourth IFOAM-Asia Scientific Conference and General Assembly in Tagaytay city (south of Manila in the province of Cavite). A seed fund for a National Certification and Inspection Body was

[8]IFOAM, established in 1972, has over 800 member organizations from 120 countries. IFOAM, sponsored initially by the developed countries, has evolved as an international organization for promoting the global organic agriculture through absorbing the developing countries.

[9]The "teikei" system is a producer–consumer copartnership within Japan Organic Agriculture Association. The ten principles of TEIKEI include: mutual assistance, intended production, accepting the produce, mutual concession in the price decision, deepening friendly relationships, self-distribution, democratic management, learning among each group, maintaining the appropriate group scale, and steady development. See "teikei" system, the producer–consumer copartnership and the Movement of the Japan Organic Agriculture Association: Country Report for the first IFOAM Asian Conference 19–22. Aug. 1993 in Hanno, Saitama, Japan. http://www.joaa.net/english/teikei.htm (Accessed August 25, 2015).

set up. From the experiences of participating in numerous IFOAM activities, the Philippines has become a self-confident and conscious inventor of organic agriculture has learnt from overseas experiences and simultaneously, creatively constructed its own organic agriculture model.

Within the Philippines, both government and non-government organizations have pushed forward the organic agriculture movement for different purposes. The organic agriculture movement sponsored by the government shifted from the governance of environmental problems resulting from industrial agriculture to stimulating organic agriculture in the country. In 1993, the Department of Agriculture, according to its commitments in "Agenda 21,"[10] established the Bureau of Soil and Water Management, and launched the Balanced Fertilization Program and the Integrated Pest Management Program. These aimed at lessening environmental damage resulting from the overuse of chemical fertilizer and pesticide, controlling the pests efficiently, improving the soil fertility, and maximizing the economic benefits based on keeping the agricultural environment healthy (Sajise and Briones 2002, p. 145). In 1998, the Agriculture and Fisheries Modernization Act was passed for pushing agricultural modernization, maintaining food health, decreasing poverty and malnutrition, inhibiting environmental degradation, and enhancing agricultural competitiveness. In 2001, the Department of Agriculture organized the Organic Certification Center of the Philippines and launched the Organic Certification and Inspection Program. According to its strict regulations, the Center rigorously inspects at least for 3–6 months before issuing the certificate that is valid only for one year. Obviously, this inspection is helpful for the export of Philippine organic produce, because it satisfactorily meets international standards. In 2002, the Bureau of Agriculture and Fisheries Product Standards promulgated the Philippine National Standard for Organic Agriculture, which is also based on international standards. Organic products certified on this standard mean that it is easier to export to the USA, EU, Japan, and South Korea.[11]

To strengthen this institutional development, the National Organic Agriculture Board (NOAB) was established based on the Executive Order 481 signed by President Gloria Macapagal Arroyo in 2005. The NOAB aims at promoting organic agriculture in the Philippines by setting up a network of organic agriculture stakeholders and creating a roadmap for further developing organic agriculture. Hereafter, the promotion of organic agriculture, as an integral part of National Agricultural Development Plan, was one of the main tasks that related branches of national government should cooperate on pushing forward. Furthermore, local

[10]Philippine Agenda 21, a national blueprint for sustainable development for the twenty-first century, was written in response to the 1992 United Nations Conference on Environment and Development in Rio de Janeiro. It envisions a better life for all Filipinos through the development of a just and economically vibrant, diverse yet cohesive society characterized by living in harmony and within the limits of the carrying capacity of nature. See: "Philippine Agenda 21," http://pcsd.neda.gov.ph/publications/pa-21/ (Accessed August 25, 2015).

[11]Carating and Tejada (2012) provide a detailed overview of the organic products that are produced and to where they are exported.

governments should actively develop small-scale organic farming based on a combination of local agricultural traditions and demands from the market.

In 2005, Congress passed "Republic Act No. 10068," known as the "Philippine Organic Agriculture Act," that mandates organic agriculture as the national agricultural strategy of the country. In line with it, in 2012, the Benigno Aquino III cabinet approved the "National Organic Agriculture Program (NOAP) 2012–2016," in which the NOAB set a target that at least 5% of the Philippines' agricultural farm areas should practice organic farming by 2016.[12] Observing these policies and activities implemented by the government, we can see how the state has constructed a comprehensive legal and management system in which organic agriculture can develop within its main framework of governmental policy.

In comparison with the national government, local NGOs have paid more attention to popularizing knowledge of local organic farming and its practice. Strengthening bottom-up participation, there are various NGOs working in urban and rural areas of the country. In 1980, the Farmer Assistance Board (a local NGO for rural development) began to criticize environmental problems that arose from the green revolution. In a critical report, "Profits from Poison," the green revolution was regarded as a form of agriculture characterized as having minimal value (Carating and Tejada 2012). In 1986, the National Farmers Convention initiated a Farmer–Scientist Partnership for Agricultural Development (MASIPAG) that encouraged farmers to plant their own varieties, apply manure, and refuse chemical inputs. In 1990, as many as 15 local NGOs, engaged in the widespread promotion of organic agriculture and organized the Sustainable Agriculture Coalition. This was devoted to holding an exhibition of organic agriculture, and teaching farmers to understand and practice organic agriculture based on lowering the external-input in agriculture (see Carating and Tejada 2012).

Many NGOs supporting organic agriculture formed their networks and umbrella governance structures with local branches working in and for rural communities. However, this did not mean that these NGOs were domestically oriented. On the contrary, they have created close contacts with international counterparts. MASIPAG is not only one of the main members of the Asian Network for Sustainable Organic Farming Technology (ANSOFT), but also members of IFOAM. These did not only introduce advanced ideas and experiences of organic agriculture to local farmers, but influenced NGOs to also set standards of organic agriculture and organic products based on a mixture of international standards, put in place by the international organic agriculture certification organization and applied in the international organic products market, and crucially within the characteristics of the Philippines' own organic agriculture. Through organizing a number of symposia on organic agriculture standards with farmer organizations, local government units, and other NGOs, MASIPAG persuaded local governments to recognize and adopt the "Participatory Guarantee System" (PGS) and further

[12]See: National Organic Agriculture Board, *The National Organic Agriculture Program* 2012–*2016* (Diliman, Quezon City, 2012), p. 11.

promote the development of organic agriculture in rural Philippines (see Bosito and Buena 2014).

Fieldwork has clarified that recently there exist different models of organic agriculture developed, which we can categorize into four types. These range from large-scale agri-commercial set-ups down to small-scale farm initiatives.

The first is the organic farm model set up by corporations for commercial agricultural production that aims at producing organic products for the national and international organic product market. Complying with the international standards of organic agriculture on soil, fertilizer, pest control, lighting, packaging, and export, this farm planted organic vegetables and fruits with high added value. During fieldwork conducted in Baguio in February 2013, a visit to the vegetable greenhouses of Golden Acre Farm (GAF), Inc. highlighted the high level of internationalization of its facilities. GAT had installed drip irrigation facilities from Israel, a yellow organic insecticide brand from India, mechanical facilities from the USA, organic fertilizer from the Philippines, a technological instructor and manager from the UK, and its products (pepper, lettuce, tomato, strawberry, celery, etc.) are exported to the markets of Tokyo, Seoul, and Europe. What was chosen to plant depended on different altitude temperatures. Soil used in the greenhouse was synthesized from various organic fibers, such as plant stems and pine fragments. Organic fertilizers (hen dung or pig fat) were chosen to apply depending on the needs of plants. What is clear here is that this organic agriculture was high-tech and capital-intensive producing high-quality products with prices that matched.

The second model is organic farming that developed out of commercial plantations. The Sycip Plantation, located in the province of Negros Oriental, owns 800 ha of land with various topographic features (mainly composed of hillsides, coastal plains, and ponds). In the framework of modern industrial agriculture, monoculture was the main production mode on this plantation. One report records that in 1982, sugarcane accounted for 86% of the total farm land and 90% of the income was derived from the sugarcane farm (Sycip et al. 2002, p. 4). Although there were other forms of economic activity such as fish and salt making, they were not efficiently integrated into the whole set up. Due to the increasing investment in changing the physical topography and applying chemical fertilizer and low productivity, the benefit–cost ratio diminished leading to the impoverishment of 2000 workers employed on the plantation. However, the most serious problem was soil erosion. All of these issues meant that the plantation was unsustainably experiencing a crisis. In order to decrease the risk of workers' grievances and develop agricultural production, the plantation had to implement a diversification policy and integrate its different sectors. In 1993, after the workers acquired 31% of ownership of property according to a Comprehensive Agrarian Reform Program, the name of the plantation changed to the Sycip Plantation Farm Workers Multipurpose Cooperative (see Sycip et al. 2002). For the workers, the conversion of ownership forced them to consider how to sustain its productivity. The workers' cooperative has practiced a new development strategy that integrates various sectors based on the diversification of planting and breeding. In this production system, raising cattle not only supplies beef, but also provides nutrition to fish and soil; planting mango trees, and fostering breeding pastures are not only helpful for the diversification of farm components but also for improving a degraded ecology and environment. The making of these chains of nutrition and biological chains have transformed the commercial plantation into an organic farm in which the economic efficiency and environmental sustainability have been guaranteed. However, although this is a large-scale organic farm, the number of this kind of plantation is still small.

The third model is an organic farm that was transformed from a small-scale family farm that had sought to increase income and secure food and nutritional demands. A typical example of this farm is Mr. Rodolfo Oray's farm in the Province of Negros Occidental. In this 10 ha farm, sugarcane, rice, and corn were once planted together. During the green revolution, the intensive input of high-yielding varieties, chemical fertilizers, pesticides, and herbicides supported by the national concessional loans produced good harvests and economic benefits. However, these successes decline in 1985. Due to an increase in inputs and a subsequent decrease in soil fertility, agricultural production was no longer sustainable, namely, income could not cover expenses needed to maintain output. As a member of MASIPAG, Mr. Oray had to convert from high-external-input modern agriculture to low-external-input organic agriculture, and further transform his family into a diversified integrated farm. According to soil structure, fertility, and topographical conditions, farm land is divided into various production areas. These areas cover lowland rice, salt, guava, and banana. Other areas cover communal forest, soil, and water conservation measures, kaingin, fallow, communal farm, fishponds, medicinal plants, nursery, and livestock components. This small but fully-functioning farm satisfied the food and nutritional demands of his family members but at the same time, has increased income through selling organic products (see Oray et al. 2002, p. 10–14). However, it needs to be pointed out that this change did not happen automatically or naturally. Change depended on the farmer acknowledging the nutritional cycle and market circulation. In other words, the farmer might get assistance from NGOs that extended local organic agriculture. In comparison with the large-scale organic farm, this small-scale venture was 50% self-sufficient and 50% commercialized, yet its amount was considerable to a large-scale one.

The fourth model is a traditional organic farm that existed widely in the Philippines (predating the Spanish colonial period). In the Cordillera Administrative Region, the Bontok have practiced their own traditional rice terrace farming system. They use indigenous seeds passed down over the centuries from their ancestors, fertilizer is mainly a compost of green manure composed of weeds, floating algae and mountain soil, and others. Pests and diseases are controlled using natural methods, such as draining paddy fields and ploughing deeply. Using these environmental-friendly technologies, rice yields in the paddy terraces have been reported to reach 6.2 t/ha (the national average yield was 2.5 t/ha) (Sajise and Briones 2002, p. 21). All products are all organic. Through such a practice, the Bontok have preserved their traditional religion and culture, which centers around terrace rice farming.

Different from the Bontok, the Ifugao have practiced an indigenous agroforestry production system, in which the forest (or Muyong) not only conserves water and soil but also provides non-timber forest produce. In the traditional swidden field (or *Uma*), rice, beans, corn, potato, tobacco, and vegetables are intercropped for 2–4 years with a 7–8 years fallow period. In the irrigated terrace (or *Payoh*) fish farming is incorporated, and wetland rice is harvested twice a year. These three pillars of agroforestry production interconnect as a whole system, in which *Muyong* conserves water for rice paddy, *Uma* contributes to biodiversity and the complete chain of food based on the conservation of water and soil, *Payoh* produces a fruitful harvest and is carried on from generation to generation (see Serrano 1990).[13] In 1995, the Ifugao terrace was listed as a world heritage site by UNESCO a designation that recognizes the organic characteristic and sustainability of the Ifugao traditional production system. Observing these two examples, we see that traditional ecological knowledge and its practice in agriculture can be reinvented and further empowered in the organic agriculture epoch. In this sense, this traditional model is the real and unique organic agriculture of the Philippines that is worth considering as a blueprint for widespread implementation.

Underpinned by two dynamics—the spread of the international organic agriculture movement in the Philippines and the reinvention of traditional Philippine agriculture—the four organic agriculture models above operate as generalizable forms that have developed in the country. Although their market orientations, management styles, and cultural basis differ, these models are quite efficient in developing a domestic organic agriculture industry. Indeed, they have been productive and remarkable in comparison to other similar practices in Southeast Asia as a whole.

Latest Measures for Promoting Organic Agriculture in the Philippines

After the signing of Philippine Organic Agriculture Act on April 6, 2010, the NOAB promulgated the National Organic Agriculture Program (NOAP) (2012–2016) in March 2012 in accordance with the Medium Term Philippine Development Plan (2011–2016) and national antipoverty strategy. The program aimed at guiding and promoting the development of organic agriculture in the Philippines. Its guidelines embody concrete organic agriculture projects carried out by various bureaus or authorities of Department of Agriculture and agencies or branches of NOAB and in their plans for the development of organic agriculture by local governments.

The NOAB (2012–2016) envisions the objectives of organic agriculture development in the Philippines. One of these objectives is to increase the ratio of organic agriculture areas in overall agriculture areas and this should reach 5% (as stipulated by the NOAB) of arable land in 2016. The second objective is to achieve recognition of the Philippine organic products by national and international consumers in terms of sustainability, competitiveness, and food security. Through increasing productivity and reducing external inputs, local farmers who are engaged in organic

[13]Cf. Sajise and Briones (2002, pp. 53–55).

agricultural production can improve their incomes to ultimately reduce poverty, protect their health, those of consumers, and the general public. Simultaneously, changes will lead to enhanced soil fertility and farm biodiversity, a reduction in pollution and destruction of the environment, as well as the prevention of further depletion of natural resources, improved resiliency to disaster risks and climate change vulnerabilities caused by human interventions and natural-induced hazards. Finally, it would meet the basic material needs and improve standards of living for all, uphold human rights, gender equality, labor standards, and the right to self-determination.[14] In comparison with industrial agriculture, organic agriculture can solve the problems in the development of the Philippines' agriculture, especially in rural areas and meet the diverse and comprehensive demands of people.

Based on the essential principles of NOAP, provincial governments are supposed to draw up their own organic agriculture programs that meet the needs of local environmental and agricultural conditions. For example, the province of Negros Occidental has proposed to make itself the first model province of organic agriculture in the country. In combination with the province of Negros Oriental, Negros Occidental has put forward the goal that Negros island should be reoriented as a "Negros Organic Island" with a vision of making it the "organic food bowl of Asia." In line with this vision, a provincial committee of organic agriculture management has been set up in Negros Occidental as well as the creation of organic agriculture ordinance and programs.[15] Furthermore, the provincial government has decided to invest 20 million pesos to push forward organic agriculture.[16] Various agricultural production units are requested to draw up their own organic agriculture programs and to strictly implement them. The programs emphasize the importance of protecting local biodiversity and farmer's agricultural practices and stimulate farmers to engage in organic agriculture. It has also established Center for Organic Agriculture Training and Extension, where it offers some courses on how to improve soil fertility using earthworms, how to choose available biopesticidal and bioherbicidal technologies, and how to successfully sell organic products in national and international markets. It has also organized the Negros Island Organic Farmer Festival to display and sell local organic products and to extend its organic products and organic agriculture experiences to the rest of the Philippines and Southeast Asia. Hence, Negros Occidental is recognized as an organic agri-tourist destination that absorbs more and more tourists through visits and exchanges. The provincial ordinance 007 was passed and bans the entry of genetically modified organisms (GMO) their planting, fertilizing, pesticiding, and herbiciding in the

[14]See National Organic Agriculture Board, *The National Organic Agriculture Program* 2012–*2016* (Diliman, Quezon City, 2012), p. 11.

[15]The Negros Occidental Provincial Board passed Organic Agriculture Ordinance of the Province of Negros Occidental, Provincial Ordinance No. 007 Series declaring a ban on living GMOs from entering the province, and Organic Agriculture program of the Province of Negros Occidental.

[16]See Danny B. Dangcalan, "P20-M fund eyed in proposed Organic Agriculture ordinance". http://www.philstar.com/region/658558/p20-m-fund-eyed-proposed-organic-agriculture-ordinance (Accessed August 25, 2015).

province of Negros Occidental. Indirectly, this ordinance has helped local farmers to promote organic agriculture. As a result of these implementations, organic farm area is now more than 5% of the total agricultural area available in Negros Occidental. This ratio is higher than the world average (3%) (see Juan 2015).

In addition, the Bureau of Agricultural Research (BAR) and Research and Development and Extension (RDE) unit of the Department of Agriculture have combined to set up the organic agriculture RDE network. This incorporates research and educational institutions, local government units, and local NGOs. In its activities, it recognizes stakeholders, such as organic fertilizer producers, retailers, agronomists, agricultural technicians, and local farmers. After intensive investigations and negotiations, the BAR put forward a unified RDE agenda and program on organic agriculture, which focuses on the innovation and application of production technology and capacity/knowledge building. In terms of technological innovation, this includes technology for production and processing of organic food, for improving and maintaining fertility, and for quickly controlling pest botanically. Technology that would guarantee high yields during the transition and conversion period from conventional agriculture to organic agriculture should be created for making the transition stable and smooth. Additionally, the production of organic fertilizer should be quickly standardized and the technology should be improved for producing good-looking and high-quality organic products. In terms of capacity building, its main aim is to extend the knowledge of organic agriculture to farmers and local officials and renew their agricultural knowledge. Importantly, it encourages them to consciously produce in accordance with (international and local) standards of organic agriculture.[17]

In its activities, a number of agriculturalists, farming engineers, NGOs activists, and officials of training center of all-level governments travel to the targeted villages and farmlands to disseminate current technological inventions and to help foment management experiences and guide local farmers.

Farmers are introduced to the fact that inputs in organic agriculture (mainly using manure and botanic pest controlling technology) is less than in industrial agriculture, while the output in organic agriculture is higher than in conventional agriculture (the premium price of organic products is usually 10–30% higher than conventionally-grown crops). This is part of an ongoing campaign to entice farmers to engage in organic agriculture based on a reasonable explanation of the huge demand and limited supply of organic products available in the market.

Some farmers may question or reject the introduction of organic agriculture based on their "received wisdom" or "conventional knowledge." Such concerns are that "organic agriculture will cause famine," or that "vegetables or fruits with pests or pest damage are organic." While it is true that there was a three-year transition-conversion (from conventional agriculture to organic one) period in which the yield

[17]See: Rita T. dela Cruz, "Strenthening organic agriculture RDE: BAR's initiatives on food sufficiency and sustainable agriculture". http://www.bar.gov.ph/organic-agriculture-rde (Accessed August 24, 2015).

is reduced. However, once this period ends, organic agriculture is ecologically more sustainable and economical in the long run. Hence, this process is explained to the farmers to convince them that the yield reduction does not necessarily lead to famine, especially when there is a problem of unequal distribution of food.

By practicing organic agriculture, with the help of technological and financial supports from both the government and market mechanisms, farmers can spend and complete the transition period smoothly through shortening the process of soil reconstruction and diversifying the planting of various varieties. Farmers notice that the organic fruit and vegetables they produced sometimes may look unattractive in shape and color, but they understand these are the main characteristics of organic foods. Recently, however, based on the comprehensive application of botanical measures of multiple cropping, intercropping, crop rotation, net bagging, netted plots, yellow sticky traps, and mulching for pest control, organic agriculture can reasonably sustain the shape of organic fruit and vegetable.[18]

Aside of these direct contacts with the local farmers, BAR also provides a space for agricultural experts to compile a handbook on organic agriculture. As a way of disseminating know-how, this manual is translated into a number of different local dialects to help local farmers understand and easily practice organic agriculture.

In order to stimulate local farmers to develop organic agriculture, local governments and NGOs have worked together to break out the orthodoxy of certification of organic agricultural products by third parties, by introducing the "Participatory Guarantee System" (PGS) widely acknowledged in the international organic agricultural community. According to the Organic Agriculture Act (Sect. 17), all organic agricultural products sold in the market should be certified by a third party. The certification fee, however, is relatively expensive (50,000 pesos a year), a price that is outside of the price range of many small-scale farmers. In 2008, the PGS was recognized by IFOAM and FAO. Under the guidance of IFOAM, MASIPAG promulgates its own PGS that is based on the international standard of organic agricultural products and mutual trust. The PGS combines various stakeholders (such as producers, consumers, local officials, scholars, and representative of NGOs) to guarantee the quality of organic products and share responsibilities.[19] PGS has reduced the standard certification fee to 700–1000 pesos for two-acre farmland and 3000 pesos for more than two-acre farmland.[20] While the standard third-party certification system mainly operates for organic agriculture that is export-oriented (for the international market), PGS serves mainly for small-scale farmers to produce organic products for local and national markets. Although PGS

[18]See Rita T. dela Cruz, "To be or not to be organic", http://www.bar.gov.ph/organic-farming (Accessed August 25, 2015).

[19]In terms of responsibilities, all stakeholders have a duty to guarantee the quality of organic products, meanwhile, each of them should bear blame and accept losses if the quality of organic products is not guaranteed.

[20]See Germelina Lacorte, "Organic food growers seek an easier way of certification." http://newsinfo.inquirer.net/632090/organic-food-growers-seek-easier-way-of-certification (Accessed August 25, 2015).

is acknowledged by IFOAM and some local governments in the country, it is yet to be acknowledged officially at a national level until the Organic Agriculture Act is amended. As the amendment of acts in the Philippines is a protracted process, some local governments, supported by local NGOs, have implemented PGS within their local boundaries.

One example of a local government that has adopted PGS as a certification for organic products is Davao city. Through adoption and setting up groups tasked to certify organic products, Davao city government has been serving its small-scale farmers. At the same time, the city government has also promoted the local economy and its sustainability. This has furthermore improved farmer's productivity through enlarging the local organic product market.[21]

Through the coordination of activities between national and local governments, local NGOs, farmers, and other stakeholders, organic agriculture has taken roots and developed relatively well in the Philippines within in a short period of time. Although there is yet a series of systematic data of organic agriculture in the Philippines, my own verification of the available statistical data confirms the development trend of organic agriculture in the country, for at least the past decade.[22] In 1999, there were only nine organic farms that accounted for 95 acres in the Philippines. A year later, the number of organic farms grew to 500, increasing to 2000 acres. In other words, 0.02% of total farming area in the Philippines. In 2009, there were 3051 organic farms with 52,546 acres that accounted for 0.45% of the total farming area. In 2013, there was a slight decline in farms at 3008 but total acreage almost doubled to 101,278 accounting for 0.8% of total farming area (see: Willer and Lernoud 2015, p. 169). These numbers alone confirm how organic agriculture in the Philippines has been growing rapidly changing fortunes for local farmers.

Conclusion

In the Philippines, agriculture is one of the main socioeconomic sectors that has been deeply affected by internal and international factors. In the context of economic and ecological unsustainability that arose through the rise of industrial agricultural practices, the Philippines' government and civil society have promoted the organic agriculture movement, based on the penetration of IFOAM and the reinvention of traditional agriculture. Over the past decade, farming areas have

[21]See Carmencita A. Carillo, "Davao farmers push gov't institutionalization of guarantee system for organic products". http://www.bworldonline.com/content.php?section=Agribusiness&title=davao-farmers-push-gov&rsquot-institutionalization-of-guarantee-system-for-organic-products&id=93162 (Accessed August 25, 2015).

[22]Some data were on the organic farms certified by the third party. Some data were from the statistics of IFOAM. All of these data did represent the organic farming that was not certified by the third party.

undergone rapid expansion to capture a rising share of the market and it is possible to observe that organic agriculture is shifting the orientation and trend of the country's agriculture.

The dramatic expansion of organic agriculture depends on the change of Philippines' economic and social structures and political will. In the present structure of the capitalist world-economy, local farmers who practice industrial agriculture will continue to fall into a "path dependence trap." The conscious development of organic agriculture arises as a response to the failures of the existing structure and offers local farmers a path that is potentially economically viable, to their own advantage and has the potential to rehabilitate degraded arable land. Although this path maybe a protracted and long process it offers a liberating alternative for farmers. How this plays out in the Philippines, may offer strong suggestions for other farmers in Southeast Asia, and other regions around the globe.

References

Anti-slavery Society. (1983). *The Philippines: Authoritarian government, multinationals and ancestral lands*. London: Anti-Slavery Society.

Bosito, E. F., & Buena, Rowena. (2014). PGS continues to gain ground in the Philippines. *Global PGS Newsletter, 05:02*(November/ December), 2–3.

Boyce, J. K. (1993). *The Philippines: The political economy of growth and impoverishment in the Marcos era*. Manila: Ateneo de Manila University Press.

Briones, N. D. (2005). Environmental sustainability issues in Philippine agriculture. *Asian Journal Agriculture and Development, 2*(1–2), 67–78.

Carating, R. B., & Tejada, S. Q. 2012. Sustainable organic farming in the Philippines: History and success stories. Paper presented at the workshop on ANSOFT-AFACI Pan-Asia project at Gwangju, Republic of Korea, 18–20 October.

Concepcion, M. B. (1994). The Philippine population problem: Myth and reality. *Philippine Studies, 42*(2), 139–154.

Corpuz, O. D. (1997). *An economic history of the Philippines*. Quezon City: University of the Philippines Press.

National Organic Agriculture Board (NOAB). (2012). *The national organic agriculture program 2012-2016*. National Organic Agriculture Board (NOAB), Department of Agriculture: Quezon City.

Oray, R., Edano, M. L. S., & Zamora, O. B. (2002). *Sustainable agriculture in a small-scale resource-limited farm: Case documentation of a MASIPAG farmer in Hinobaan, Negros Occidental*. Los Banos: SEAMEO SEARCA.

Sajise, P. E., & Briones, N. D. (2002). *Environmentally sustainable rural and agricultural development strategies in the Philippines: Lessons from six case studies*. Los Banos: SEAMEO SEARCA.

Serrano, R. C. (1990). *Environmental and socio-economic impact analysis of an indigenous and an introduced agroforestry system in Luzon*. Unpublished PhD Dissertation, University of the Philippines, Los Banos.

Sycip, M. L., Javier, C. F., Salayog, F. A., & Vilar, N. C. (2002). *Sustainable agriculture in a large-scale commercial farm: Case documentation of the Sycip plantation farm workers multipurpose cooperative in Manjuyod, Negros Oriental*. Los Banos: SEAMEO SEARCA.

Veneracion, J. B. (2000). *Philippine agriculture during the Spanish regime*. Quezon City: College of Social Sciences and Philosophy.

Willer, H., & Lernoud, J. (Eds.). (2015). *The world of organic agriculture: statistics and emerging trends 2015*. Frick and Bonn: Research Institute of Organic Agriculture (FiBL) and IFOAM-Organics International.

World Bank. (1989). *Philippines: Toward sustaining the economic recovery*, Report No. 7438-PH (January 30).

Online sources

Carillo, C. A. (2015). Davao farmers push government institutionalization of guarantee system for organic products. http://www.bworldonline.com/content.php?section=Agribusiness&title=davao-farmers-push-gov&rsquot-institutionalization-of-guarantee-system-for-organic-products&id=93162. Accessed: 25. Aug. 2015.

Cruz, R. T. D. (2015). Strenthening organic agriculture RDE: BAR's initiatives on food sufficiency and sustainable agriculture. http://www.bar.gov.ph/organic-agriculture-rde. Accessed: 24. Aug. 2015.

Cruz, R. T. D. (2015). To be or not to be organic. http://www.bar.gov.ph/organic-farming. Accessed: 25. Aug. 2015.

Dangcalan, D. B. (2015). P 20-M fund eyed in proposed Organic Agriculture ordinance. http://www.philstar.com/region/658558/p20-m-fund-eyed-proposed-organic-agriculture-ordinance. Accessed: 25. Aug. 2015.

IFOAM-Organics International. (2015). Definition of organic agriculture. http://www.ifoam.bio/en/organic-landmarks/definition-organic-agriculture. Accessed: 2. Feb. 2015.

Juan, C. (2015). Is Negros Island the organic bowl of Asia? http://www.manilatimes.net/is-negros-island-the-organic-bowl-of-asia/199984/. Accessed: 26. July. 2016.

Lacorte, G. (2015). Organic food growers seek easier way of certification. http://newsinfo.inquirer.net/632090/organic-food-growers-seek-easier-way-of-certification. Accessed: 25. Aug. 2015.

N.A. Organic farming: the future of Philippine agriculture. (2012). *Far Eastern Agriculture (FEA)*. http://www.fareasternagriculture.com/crops/agriculture/organic-farming-the-future-of-philippine-agriculture. Accessed: 25. Aug. 2015.

Republic of the Philippines. (2015). Act No. 10068 (known as Organic Agriculture Act of 2010). http://www.da.gov.ph/index.php/laws-issuances. Accessed: 2. Feb. 2015.

Chapter 10
Livelihood After the Dams: Experiences of Tributary Dams in the Mekong River

Yuka Kiguchi

Abstract Over the past 20 years, Southeast Asia has seen the intensification of dam-building projects across and within nations. This chapter exams the construction of dams along Mekong tributaries and their impact on the livelihood of local communities. It focuses closely on the Pak Mun Dam case in northeastern Thailand to show how local communities have been left to bear the negative impacts of dam construction, and how these have affected both fisheries and local knowledge. This chapter argues that the Pak Mun Dam case can serve as a departure point to understand the socio-environmental changes that will arise with the construction of a new dam in the region, the Lower Sesan 2 Dam in Cambodia. Ultimately, this chapter shows what is at stake in the construction of both the dams in terms of impacts upon fish diversity, the loss of fisheries in the Mekong River, and human displacement.

Keyword Mekong river · Hydropower dams · Northeast Thailand
Cambodia · Livelihood · Pak Mun Dam · Lower Sesan 2 Dam

Introduction

As an international waterway that begins in the Tibetan Plateau and flows through the Mekong Delta into the South China Sea, the Mekong River is rich and second only to the Amazon in terms of fish biodiversity (Baran 2010).[1] Along with 850 species of fish, there are 20,000 species of plants, 430 animal species, 1200 bird species, and 800 reptile and amphibian species that live in the river (Thompson

[1]According to the data registered as of 2009, it has been confirmed that the 781 fish species found in the Mekong River are second only to the 1271 species of the Amazon River, showing that the Mekong River is second in the world for fish diversity.

Y. Kiguchi (✉)
Mekong Watch, Tokyo, Japan
e-mail: kiguchi@mekongwatch.org

M. Lopez and J. Suryomenggolo (eds.), *Environmental Resources Use and Challenges in Contemporary Southeast Asia*, Asia in Transition 7,
https://doi.org/10.1007/978-981-10-8881-0_10

2008). The entire catchment area covers 795,000 km^2 and the river has a length of about 4900 km (MRC 2007).

At present along the Mekong River, there are about 152 dam plans and 52 dams under construction[2] whose main purposes are to generate electricity and for irrigation. Hydropower dam projects in the Mekong Region have been a source of controversy since the 1990s due to the number of negative impacts upon riverine ecosystems and the detrimental effects upon fish and their life cycles. The Mekong River Commission (MRC), an inter-governmental river basin organization established by country governments of the riparian zone,[3] has pointed out that the Mekong's role as a source of food is crucial with freshwater fish forming one of the most important sources of protein for about 56 millions of people (MRC 2007). Furthermore, fisheries in the region are a major source of cash income and employment for many local people.[4]

However, after more than two decades of experience with a deteriorated river ecosystem due to dam construction and operation, not one single project has been canceled on the grounds of ecological considerations. Based on fieldwork observations and interviews, fish stocks continue to decrease, however to date, no cumulative impact assessments have been carried out. It may be too late to verify the real negative impacts of the dams in a quantitative way nonetheless, in the absence of scientific evidence, fishermen have noticed, in their daily activities, that fish populations have been drastically decreasing.

This chapter focuses on the current fishery situation and the impact of the dam construction in the Mekong River by focusing on the contested case of the Pak Mun Dam in Thailand that was constructed between 1990 and 1994. Previous research on Pak Mun Dam has noted that there were "unresolved controversies and risks" since the very beginning of the project (Foran and Manorom 2009). This chapter will show how 25 years after its construction the dam has affected the livelihood of around 60 communities. To highlight the effects wrought upon fishing knowledge and skills, it focuses on one village in the affected area to illustrate what changes have occurred. The chapter is organized as follows. First, it starts with an overview of the geographic features of the Mekong River and the diversity of its fish that have shaped community life in northeastern Thailand. Based on field surveys and interviews between late 2004 and late 2015, it tracks fishermen's grounded knowledge of the river's geography and fishing activities in the Mun River, an important tributary of the Mekong River in northeastern Thailand. It also discusses the Lower Sesan 2 Dam (in Cambodia) to highlight the issue of dam construction

[2]The CGIAR Research Program on Water, Land and Ecosystems. https://wle-mekong.cgiar.org/maps/ (accessed on August 22, 2016).

[3]Countries in the riparian zone are Cambodia, Lao PDR, Thailand, and Vietnam.

[4]For example, in the vicinity of the Khone Falls in southern Laos, the Mekong River mainstream fishing industry supports more than 65,000 households. The average household in this area is thought to catch an annual average of 355 kg of fish, and consume 249 kg of fish. The total catch in the area close to the Khone Falls is estimated at 4000 tons, worth between US$ 450,000 and US$ 1 million (Baran et al. 2007).

for the local community. The Pak Mun Dam serves as a typical development project found in many developing countries that not only fails to generate income and empower the community as it has promised to deliver, but also contributes to a loss of local knowledge.

The Mekong River Basin: geography and diversity of fish

The Mekong River passes through six different countries: China (Qinghai and Yunnan Provinces), Myanmar, Lao PDR, Thailand, Cambodia, and Vietnam, before flowing into the sea. After flowing through China, the Mekong River enters Lao PDR, forming the border between Myanmar and Lao PDR (Laos), and after flowing along the Thai–Lao border, enters Cambodia to connect to the Tonle Sap Lake. The river then enters Vietnam, where it is known as the "Nine Dragons River". The Mekong Delta in Vietnam is one of the world's largest rice-producing areas, and from there the Mekong flows into the South China Sea. The Mekong originates with the melting snow of the Himalayas, but water from the river's catchment area in China makes up 16% of the river's flow, while Myanmar accounts for 2%, Lao PDR 35%, Thailand 18%, Cambodia 18%, and Vietnam 11% (Table 10.1 MRC 2005).

The lower Mekong River area is in a tropical monsoon climate zone and from mid-May to the end of September it is under the influence of the southwestern monsoon. The months between October and May experience a dry season under the influence of the northeastern monsoon (MRC 2010). In a tropical monsoon climate zone, there is a great difference in rainfall between the dry and rainy seasons which greatly affects the water level of rivers over the year.

A strategic environmental assessment order by the Mekong River Commission (MRC) has noted that there are 850 freshwater species, and that around 1100 species inhabit the river if marine fish that make temporary incursions into the Mekong River delta are also included (MRC 2010). Its tributaries are also rich in aquatic life and more than 200 species of fish can be found in the Songkhram River basin and the Mun/Chi basin in Thailand as well as in the Sekong and Srepok basins in Cambodia. It is, therefore, not a surprise that fish and other freshwater aquatic species are the main sources of protein for the people who live in rural areas

Table 10.1 Comparison of the Six Countries in the Mekong River Catchment Area

	China	Myanmar	Lao PDR	Thailand	Cambodia	Vietnam	Entire Region
Catchment area (km^2)	165,000	24,000	202,000	184,000	155,000	65,000	795,000
Catchment (% of entire watershed)	21	3	25	23	20	8	100
Volume (% of entire watershed)	16	2	35	18	18	11	100

Source MRC 2005, 1

Table 10.2 Species Richness in 20 Locations of the Mekong Basin. *Source* Baran 2010, p. 10

Location	**Species**	**Families**	**Endemic**
Mekong Mainstream			
China-headwater	24	3	4
China-upper reach	34	4	4
China-middle reach	48	8	7
China-lower reach	122	21	15
Northern Lao PDR	140	30	26
Mekong down Khone Falls	168	34	25
Stung Treng-Kratie	204	37	33
Mekong Delta	486	73	28
Mekong Tributaries			
Nam Ou	72	15	29
Nam Ngum	156	27	43
Nam Mang	57	19	17
Nam Kadinh	99	21	38
Songkhram	216	40	39
Xe Bang Fai	157	31	51
Xe Bang Hiang	160	33	47
Mun/Chi	270	38	49
Sekong	214	33	63
Sesan	133	26	24
Srepok	204	32	38
Tonle Sap Lake	284	45	31

of the lower Mekong basin (the four countries of the lower basin, not including China and Myanmar). Fish species diversity in 20 locations in the river basin (including tributary river sub-basins) are shown in the following table (Table 10.2).

It is important to note that the migration of fish populations is an important element in their life cycle, especially in terms of spawning and growth. In the Mekong River basin many fish migrate between the Mekong's mainstream and its tributaries against very diverse conditions during both the dry and rainy seasons. Three distinct, but interconnected, migration patterns have been identified in the lower Mekong River basin, each involving multiple species. According to one study by MRC, these are, respectively, the lower (LMS), the middle (MMS), and the upper (UMS) Mekong migration systems (MRC. 2002) (See Fig. 10.1).

Another study by AOP and SEARIN, based on a survey led by the affected community of the Pak Mun Dam in northeastern Thailand (which is located in the MMS area), clarified that migration can be observed every month (with the exception of January). It also notes that there are variations in fish species and direction, either upstream or downstream (AOP and SEARIN 2002a). The first peak migration season is through March to April, when small fish swim up the Mekong

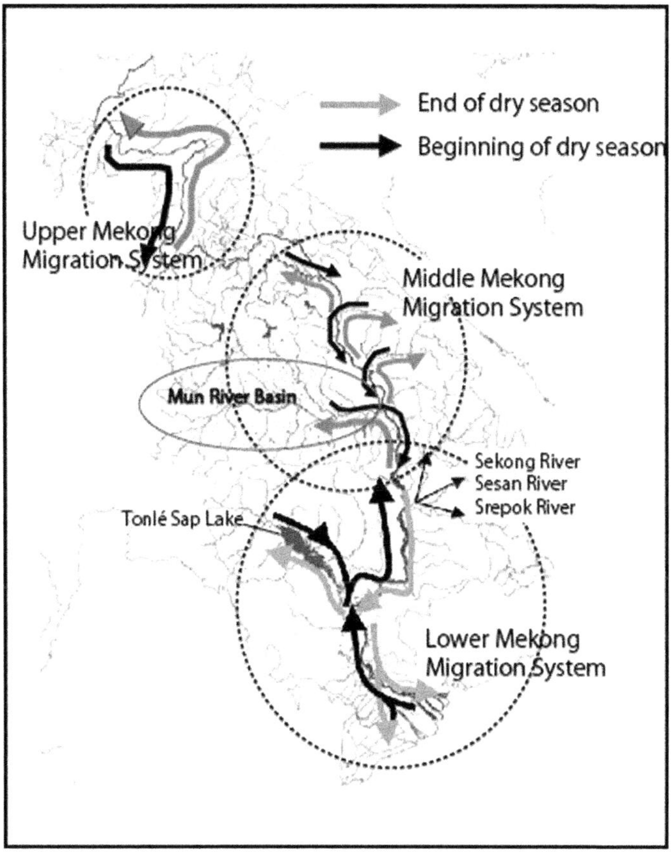

Fig. 10.1 Fish Migration Systems of the Mekong River. *Source* Mekong River Commission (MRC) 2002

and into the tributaries such as the Mun River. The next season is from May to June, when carnivorous migratory fish swim up the Mekong and also enter tributaries and floodplains. Between October and December, fish that migrated upstream begin to return downstream into the Mekong mainstream (See: AOP and SEARIN 2002a, b). These kinds of migration patterns can be seen in the tributaries in the left side of the Mekong River in Laos and the northeastern Cambodia.

This rich diversity of fish species in the Mekong River is an important characteristic of the region and in itself, has shaped the fishery contours there, especially in the lower Mekong River which is experiencing issues of productivity due to overfishing and subsequent declines in stocks. According to Baran (Baran 2010, p. 12) the amount of freshwater fish catches in the four countries in the lower Mekong is around 750,000 tons per year. However, from field surveys data, estimates suggest that fish catches could be around 2.1 million tons per year, which is equivalent to 18% of all global freshwater fish catches (Baran 2010, p. 12). For Mekong Basin fisheries, these are estimated to have an annual economic value of between US$ 2.1 ~ 3.8 billion and $4.2 ~ $7.6 billion at retail prices (Baran 2010).

Different sources estimate that freshwater fish production in Cambodia, including fish farming, accounts for between 11.7% and 16% or between 8% and 12% of GDP. While these figures are all estimates, and it is difficult to obtain accurate statistics for small-scale fishermen. Any importance given to fisheries can only be understood from these macro-level approximations (Baran 2010).

The Mun River Basin in northeastern Thailand

> The Mun River is like a miraculous rice field. In this field, thousands of people plant rice at the same time and then harvest it. On this river, even if many people fished at the same place, they could all catch fish, so there was never any need to stop anyone from doing anything. In addition, different from rice fields on land, those who would come first cannot claim possession. The river was everyone's asset (Fisher folk from Khongjiam District, Ubon Ratchathani) (Mekong Watch 2004).

One of the main tributaries of the Mekong is the Mun River that flows through northeast Thailand, and the area of its basin is the largest of all Mekong tributaries in Thailand (See Fig. 10.2). The catchment area of the river is 117,000 m^2 (TEAM Consulting Engineers Co., Ltd. 1982). The source of this 750 km long river is in Nakhon Ratchasima Province, and it flows into the Mekong River at the Thai border with Laos in Ubon Ratchathani Province. The riverbed in the downstream area of the Mun River possesses a complex topography consisting of rapids and pools, and this continues upstream for 30 km from the mouth of the river. The environment offers an appropriate place for nurturing a variety of fish species (AOP and SEARIN 2002a, b). Due to a lack of suitable land for agriculture, fishing villages have formed and grown along the banks of the Mun River (Kanokwan 2004). In the middle Mun River Basin, there are seasonal wetlands—complexes composed of many oxbow lakes and flood forests called "*par bung par thaam*" (wetland and flood plains). These areas become connected to rivers during the rainy season, functioning as fish habitats and their spawning and nursery grounds (Research Committee of the Villagers of Rasi Salai 2004). The migration of many freshwater fish can be observed along the Mun River. With these rich aquatic resources, fishing has became a main source of livelihood for villages located within the community along the lower Mun River (AOP and SEARIN 2002a, b; Kiguchi 2006).

According to local knowledge of the fishing villagers in this river basin, there are three peak seasons for fishery. The first is from February until April and is the season when small fish from the Mekong River migrate to the Mun River. The second is between June and July a season for other kinds of fish, mainly carnivorous types that migrate into it. The last is during October and December, when the water level declines and fish migrate to the Mekong (AOP and SEARIN 2002b; Mekong Watch 2004.).

Based on this practical knowledge, villagers categorize topography and the use of fishing gear dividing the river's topographical features into 22 types (AOP and SEARIN 2002b). Based on their knowledge along the lower Mun River, Ubon Ratchathani Province, the topographical features of the surrounding area are divided the following way (see Table 10.3).

A number of studies have noted that in the lower Mekong River basin, the fishing grounds established by Lao ethnic villagers (known as the "*Luang*") have been long established by local people (Kanokwan 2004; AOP and SEARIN 2002a; Kiguchi 2006). *Luang* is a combination of merely a place for fishing gear installations and a mixture of management and fishing methods (see Table 10.4). Given the fact that local communities possess their own local knowledge on the topographical features of the river and various fishing gear, they have also developed a complex set of rules regarding the usage of these. Whether customary fishing rights—a traditional system employed in the area—are applied or not very much depends on the fishing gear.[5] Every household continuously maintains customary fishing rights in *luang* and some are individually used while others are used by multiple persons (Kiguchi 2006).

Toum yai holds great importance in the daily practice of the local community. As a fish trap, it has the shape of a cone with a hemisphere attached to its bottom (see Fig. 10.3).[6] It is approximately 7 m long and more than 1 m in diameter at the bottom. It is set in the areas around K Village in Phibun Mansahan District, Ubon Ratchathani Province, and only at the mouth of the Mun River. K Village is a Lao village, with limited land allocated to rice paddies. In 2005, approximately 50% of the villagers were growing rice (Kiguchi 2006).

Surveys from the mid-2000s have revealed that *toum yai* fishing started in this area after Vietnamese refugees from the Indochina War immigrated there (1978–1979). Fishing methods were later developed by the Lao people who learned how to use fishing tools and expand their markets. The main means of livelihood in this village was fishery and cash income earned mainly by day labor and fishing. For fishing, villagers predominantly used gill nets and long lines, but also used *toum yai* during the rainy season. In some ways, *toum yai* has become an important source of income to buy rice for the people in K Village who do not own paddy fields (Kiguchi 2006, 2014).

Toum yai is also called "*toum par yon*" because it is mainly used to catch *Pangasius macronema* (locally known as *par yon*). *Par yon* is said to migrate between the Mekong mainstream and its tributaries. They come up the Mun River in May and June when the rainy season starts and rice is used for bait. The local community has developed knowledge of the river, studied *par yon*'s behavior and improved the shapes and usage of fishing traps. This has given rise to a unique

[5]Villagers who own the customary rights of using *Luang* are known as *Cao Khong* (master, owner), however those people do not really own a place, but only the right to exclude other people from using it. Therefore, without permission from *Cao Khong*, other households cannot freely use *Luang*.

[6]*Toum* means "a fishing trap made from bamboo" and *yai* "big".

Fig. 10.2 Map of the Mun River

Table 10.3 Topographical features classified by villagers. *Source* AOP and SEARIN 2002b and the author survey in 2005

Local Names	Meaning (in English)	Features
Do'o'ng	Island	Island in the river
Hin	Rock	A rock that is exposed above the surface of the water during the dry season
Kaeng	Rapid	Shallow, fast flowing part, rocks on the riverbed
Khan (Khan jo'm, Khan tu'u'n)	Flat Rapid (Deep flat rapid, Shallow flat rapid)	Rocks on the riverbed, but at ***Khan*** they are not exposed above the surface of the water even during the dry season. Deep and shallow ***Khan*** are divided
Kho'o'n	Sandbank	Sandy soils formed downstream from rapid currents
Khum	Deep pool	Deep part or deep pool area. The riverbed could be either sand or rock
Paew	Channel	Area where ditch-like channels have been eroded into the rocks of the ***kaeng***
Phaa	Cliff	Cliff inside the river
Saang	Fountain	Spring water at the river banks
Tham	Cave/Hollow	A large crevice found in layers of flat rocks or hollows on the rocks
Woeoeng	Deep pool	A pool in which water as if a whirlpool

Table 10.4 Usage of Fishing Gears consist *Luang*

Local Names	English names (see Claridge 1997)	Features	Usage Methods and Users	Location of Use
Ii-long (Lop)	Longitudinal Cylinder Trap	Made of bamboo and nylon net. There is a spike called ***Ga*** inside to prevent fish leaving	Placed where fish pass by and bait is not used. Used by one adult man	Near the river bank
Kuat (Uang)	Beach Seine/ Haul Seine	Large net that has floats and weights	One of the net edges is fixed on the bank and the other edge is drawn over into the river by boat. Bait is not used. Used by 2–10 adult men	Sandy Riverbed
Lop [non]	Horizontal Cylinder Trap	Bamboo trap. Two spikes called ***Ga*** are installed inside to prevent fish from going out. ***Non*** means horizontal	Placed where fish pass by and bait is not used. The fence is built along to guide fish. Used by one adult man	Near the river banks
Mong	Gill Net	Nylon-knitted rectangular net. Sizes vary	Kept balanced on the water surface by floats and weights. The net is set up from the water surface facing vertically down to catch passing fish. No bait is used. Used by 1–2 adult men	Normal current, pool, tributary river
Toum yai (toum pa yon)	Bamboo Basket Trap	Large-sized cone-shaped fishing gear. Full length is 7–8 m. Made from bamboo	Rice is used as bait. *Pangasius macronema* is the main kind of fish caught with this gear. Used by 1–2 adult men or women	Pool

management system of fishing grounds. Other cone-shaped fishing traps have a narrow path woven with bamboo strings at the bottom (*gaa*) so that fish can enter but not exit. Fishers need to place pillars in the river before setting *toum yai* and these are made from bamboo longer than *toum yai*. Fishers then drive nails into them and tie rocks using the nails as hooks. Five pillars are required for one *toum yai* (Kiguchi 2006, 2014).

Toum yai's *luang* is ideally shallower than 15 m and sandy at the bottom. Deep waters with gentle water flow, called *khum,* are often chosen for *luang* of *toum yai.* Since the conditions of river beds change every year with the water flow, *luang* moves around accordingly. Joint owners of *luang* try to work the same hours to avoid disturbing each other and when bathing in the river, they do so downstream of *luang*. Due to *toum yai's* open mouth structure, fish can run away from *toum yai,*

Fig. 10.3 Toum yai

thus fishers need to work at the same time. When they moved their *toum yai* to another part of the same *luang* or to a different one, they make sure they leave enough space between other people's *toum yai* and theirs so that others can reach them by boat. (Kiguchi 2006, 2014)

Fishers pull up *toum yai* early in the morning between 3 and 5:00 am and usually work in pairs. After anchoring their boat next to a pillar, they close the bottom lid of *toum yai* by pulling a string called *saay tor* tied to the lid (see Fig. 10.4). They also close the top lid so that they can tilt *toum yai* sideways with fish inside. Then they untie the string holding the pillars and *toum yai* together, remove them from their sets and pull them up to lay them on a boat. One person operates the boat and the other person pulls up the *toum yai*. The pairs are most often a married couple and the wife is the one who operates the boat while the husband sets and pulls up the *toum yai*. Male partners usually have to dive into the river to carry this out (Kiguchi 2006, 2014).

Villagers used to catch 30–40 kg of fish per day by *toum yai* fishing, an important source of cash income which turned out to be at least 10,000–20,000 Baht (US$ 250–500) by the end of their 1–2 month fishing season in 1990s. Families who were not landowners or growing rice used to buy rice with the income they earned this way. Some families used to buy even water buffalos and other livestock (Kiguchi 2006, 2014).

Fig. 10.4 Saay tor

Until 50 years ago, the exchange of fish and rice took place over an extensive area, including southern Lao PDR and northeastern Thailand (Kiguchi 2006). In an era when subsistence lifestyles were virtually universal, people spent much of their time gathering in order to secure food supplies. In villages where the main vocation was farming, it was difficult to find time to obtain supplementary foods (to eat with rice) during the busy farming seasons of rice seedling transplantation and rice harvesting. At the same time, in riverside villages, where fish were available in abundance, they were caught and processed into fermented or dried foods, which the villagers then actively exchanged for the goods they needed. Food was thus distributed across the region through an established system of bartering (AOP and SEARIN 2002a; Kiguchi 2006).

These exchanges not only had practical aspects, but also strong social and cultural significance. A woman in her 70s living in K Village relates that in her youth she would make large amounts of dried and fermented foods, load them onto an ox cart and often set off together with a friend without first deciding on a destination. She would then negotiate with people she met along the way to exchange the fish products she had brought with her for rice and other agricultural products. There was no fixed exchange rate for rice and fish, so when she had a surplus she would give more to the other person, and if she was in short supply she would negotiate with the other person to obtain larger amounts of rice. It was understood that what was important in these negotiations was to be open handed. Interviews also clarified that setting off to an unknown village, engaging in exchanges, and making new friends with people there was considered something enjoyable to be looked forward to. Within this system of bartering, people who became friends through exchange frequently exchanged visits with each other (Kiguchi 2006).

These exchanges still exist in some communities in northeastern Thailand. According to one interview in 2013 at Khongjiam District in Ubon Ratchathani Province, some fishermen still exchanged fish and rice with their relatives who cultivated rice. They explained that they could get more rice than they sell fish and buy rice in the local market. People who cultivate rice are busy during the cultivation season, so they need a side dish that is easy to cook. Fish is the most favorite one for them. Due to this, there is a high demand for fish, so fishermen have better opportunities to procure more rice than if they trade in the market.[7]

Twentieth-Century Dam Project: The Pak Mun Dam in the northeastern Thailand

In 1994, the construction of the Pak Mun Hydropower Dam at a location of 5 km from the river's mouth blocked fish migration and severely impacted local fishing (Fig. 10.5, 6, AOP & SEARIN 2002a, b; WCD 2000; Ubon Ratchathani University 2002). The World Bank provided loan aid for the Pak Mun Dam Construction Project. The main purpose of this so-called "multipurpose dam," standing at 17 m in height and 300 m in length with an installed capacity of 136 MW, was for generating electricity (WCD 2000).

The dam was met by intense opposition not only from local communities, but also from a number of international and Thai NGOs. In December 1991, the World Bank approved the financing of Thailand's Third Power System Development Project, which included the construction costs for the Pak Mun Dam (WCD 2000). These costs exceeded the originally planned estimates and led to a significant displacement of households (Table 10.5). As suspected by NGOs and academics, the dam has had a significant negative impact on both the natural and social environment, destroying local fisheries, and leaving local communities impoverished. Before the construction of the dam, only a few people opposed it,[8] as government authorities suggested that if it was built the villagers could catch more fish. Some of them believed this, while others did not, but fear of government officers meant they did not have the courage to oppose it. After commencement of operations, many fishermen had little doubts as to the negative impacts on their fishery. This subsequently led to the formation of a people's movement to oppose the Pak Mun Dam (Kiguchi 2006). Although there was compensation to those who were considered most affected, it was only available to those who were forced to move their houses from the construction site. Those most affected only received

[7]Interview with villagers in Khonjiam District, November 2013.

[8]Interview with villagers who opposed the dam in 1999.

Fig. 10.5 Closed gate

Fig. 10.6 Opened gates

Table 10.5 Planning versus reality of the Pak Mun Dam compensation and mitigation plan

	Resettlement costs(baht)	Displacement	Agricultural land inundated	Mitigation project for fish(baht)
Planned	394.77 million (1988)	248 households (1985)	–	–
Actual	1113.1 million (1999)	943 households (1994)	706 households (1994)	305.55 million (1995–1998)

Source Pak Mun Dam, Mekong River Basin, Thailand: A WCD Case Study (2000)

compensation after they had strongly opposed the dam and negotiated with the government and the project owner, Electricity Generating Authority of Thailand.[9] However, compensation was far from the reality of the negative effects they had to bear and amounts awarded with regard to the dam's construction were reduced by nearly three times their original estimates.

In 1999, a people's movement calling for the opening of the dams' sluice gates gained strong momentum, and in 2001 the Thai government announced their experimental opening. Opened between June 2001 and November 2002, the partial recovery of plants on the banks of the river during the dry season and the return of migrating fish was observed, signaling the potential to restore the pre-construction conditions of the environment (Kiguchi 2006). A survey conducted by researchers at Ubon Ratchathani University in the affected area of the Pak Mun Dam in 2001 found 184 species from 44 fish families (Ubon Ratchathani University 2002), while villagers documented 156 species (counted according to the number of local names for the fish) (AOP & SEARIN 2002a, b). What is clear is that the Pak Mun Dam shut the migrating route of fish from the Mekong (AOP & SEARIN 2002a, b).

However, at the same time, the Rasi Salai Dam was constructed. This one dramatically changed the hydrology of oxbow lakes and flood forests, which were crucial habitats and spawning grounds for fish in the middle of the Mun River (Research Committee 2004). These two dams have had enormous negative impacts on the Mun River environment and as a result, fish numbers have drastically decreased. While this was happening, villagers continued their protests against the dam even after operations began.

Surveys also clarified that the average fishery income of local people decreased sharply from 25,742 baht before dam construction to 3045 baht afterward. These then reportedly recovered to 10,025 baht in the survey conducted when the dam gates were opened (Ubon Ratchathani University 2002). The university research team proposed keeping the sluice gates open year round on a trial basis for 5 years, citing figures on the degree of the villagers' impoverishment and the lack of any economic merits of the dam's electricity production. They documented the recovery of fishery resources through opening the sluice gates, and noted that they would

[9] Interview with affected people of the Pak Mun Dam during research, 2001.

have no impact on the electric power supply, but significantly improve economic conditions for impoverished people in villages (Ubon Ratchathani University 2002). The Thai government, however, did not adopt the proposal and decided to open the sluice gates for only 4 months of each year[10] (Kiguchi 2006).

As mentioned above, *toum yai* fishing was mainly done in K Village and neighboring villages. However, due to the construction of the Pak Mun Dam, *toum yai* fishing has undergone significant changes in order to adapt. It is important to examine the usage of *toum yai* after the dam was constructed so as to understand the contours of these changes in local people's daily lives.

In order to grasp changes in fishery, the number of *toum yai* was surveyed through questionnaires and field interviews. Each household was asked how long they were involved in *toum yai* fishing. Villagers remember the period well because *toum yai* fishing holds significant value in their economy. Answers were confirmed by relating the period to the villagers' children's age or the significant events such as marriage. Villagers were asked the names of the family members who were fishing and the number of years fishing. In spite of the fact that the wording of the question was "which year," many answered they stopped fishing "after the Pak Mun Dam was constructed." In later interviews, it was confirmed that they were not referring to 1991, when the construction started, but 1994, when its operation started with the sluice gates closing. This was the time when many villagers abandoned fishing. For this reason, when villagers answered "after the construction of the Pak Mun Dam," 1994 is taken as the point when fishing practices stopped. Answers were also validated by asking village chiefs, members of the Tambon Administration Organization and their families whether those villagers were really fishing in the period they claimed. When the period was not confirmed, data was amended by re-interviewing the villagers. The results of this survey are shown in Fig. 10.7 (Kiguchi 2006, 2014).

The figures demonstrate that most households stopped fishing in 1994 and that villagers stated virtually no fishing was done between 1995 and 2001 though some had set *toum yai* just to see if they could catch something. During this period, more than half of the villagers in K Village participated in the movement to demand compensation and open the sluice gates. These were opened on a trial base for the purpose of research in 2001. After that, some of the Mun River's fishery resources recovered and villagers resumed fishery. Villagers fished with *toum yai* again in K Village in June 2002 for the first time after 1994. (Kiguchi 2006, 2014)

The villagers' movement to open the sluice gates continued from 1999 to 2002 and responding to this, the government compromised and gave an order to open them 4 months per year from 2003. However, since the opening did not coincide with fish migrations (the peak is at the end of May to June), the catch was not large that year. Again, fishers abandoned *toum yai* fishing.

[10]Gates were opened between July and October. This decision was canceled after the collapse of Thaksin Shinawatra government with a military coup d'etat in 2006. The schedule to reopen the dam gate is uncertain. In 2015, the dam's gate was opened between August and October.

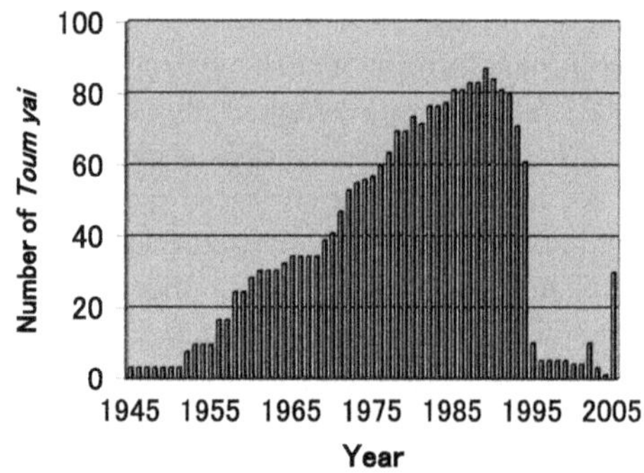

Fig. 10.7 Number of *toum yai* used in K Village

Observations during fieldwork in October 2015, 25 years after the construction of Pak Mun Dam, clarified that *toum yai* fishing has been completely abandoned. In K Village, a 50 something-year-old Mr. Somboon (a pseudonym) recalled that he has not had the chance to show off his skills of using *toum yai* in the past years. When the Pak Mun Dam sluice gates opening is delayed, there are hardly any fish to catch upstream of the dam. A few fishers tried *toum yai* fishery in 2015, but stopped after a few days as they realized that there was not much to fish. Somboon also prepared his *toum yai*, but did not have the chance to use it. He explained that it is "best is when the dam gate opens between April and May. August is too late for fish to come up the Mun river."[11] He also recognizes the significance of what is disappearing in the village when he said that, "if the World Bank had not financed the project, the Pak Mun Dam would not have been built. Twenty-five years is enough for a baby to grow up into adulthood. Youngsters in the villages do not know the river or fish. The skill for *toum yai* is on its way to becoming lost."

Similar experiences have been noted in other surrounding villages. A 32-year-old fisherman Mr. Viriya (a pseudonym) whose village was also affected with the construction of Pak Mun Dam said "If the dam gates were closed until late August again, most fish will not be able to migrate back to the Mun River from the Mekong."[12] Viriya and his father have many kinds of fishing gears and just in front of his house, they hang some made of gourd, which is rarely seen today (it is predominately used to catch fish near the surface by floating it on the rapids with hooks and baits). In 2015, he was not able to use it since the dam gate opened too late and there were no fish to catch.

After the dam's construction, Viriya had to stop fishing, which he had been doing with his parents since he was young. After graduating from school he found a job at a factory, however he suffered a severe injury in a traffic accident. Due to this, he could not continue work and returned to his village. He said that he can catch fish if he works hard, however, there is barely anyone in his generation who remained in his village to do fishing, because the fishery alone can no longer

---

[11]Interview with Mr. Somboon, October, 29, 2015 in K Village.

[12]Interview with Mr. Viriya, October 29, 2015 in D Village.

provide a stable livelihood. In 2015, there are few people who earn their living by fishing alone, especially in the lower Mun River. In the absence of opportunities to use their fishing gears, fishermen in affected villages are no longer using them and are forced to abandon fishing all together and change occupation in order to sustain their livelihood. Local knowledge and fishing skills are in danger, and so is their livelihood as fisherman in the lower Mun River.

Twenty-First Century Dam Project: The Lower Sesan 2 Dam in Cambodia

> You see, rich people in the town happy with the dam and they can use electricity. But for villagers living along a river like us? I can only think that they are trying to kill us. They are waging war against us (Comment of a woman in Andoung Meas District, Ratanakiri Province during the interview in September 2015).

The Sekong, Sesan, and Srepok Rivers form one part of the largest tributary group of the Mekong River.[13] The downstream region is made up of Cambodia's northeastern provinces of Ratanakiri, Steung Treng, and Mondulkiri. After the convergence of the Srepok and Sesan Rivers, the Sekong River then flows into the mainstream and then flows into the Mekong. The three rivers are also important spawning grounds for Mekong River fish that migrate within the river system and 89 species of migratory fish swim up the three rivers to return to the tributary group (Baran et al. 2013).

The Lower Sesan 2 (LS2) Hydropower Project is located on the Sesan River in Sesan District, Stung Treng Province, Cambodia, 1.5 km downstream from its confluence with the Srepok River and 25 km from where the two rivers meet the Mekong River mainstream. The project was approved by Cambodia's Cabinet in November 2012 (3SPN and Mekong Watch 2015). Construction of LS2 began in February 2014 and as of October 2015, the construction was being carried out by blocking the flow of the Sesan River.[14]

Located on very flat land, the 75 m high LS2 will have a 33,560 ha reservoir. According to the EIA in 2008, the project will displace 4785 villagers in 1059 households from 7 villages in 4 communes (3SPN and Mekong Watch 2015). A report, published in 2009 pointed out that upstream impacts might be much more serious than estimated in the EIA, with at least 78,000 villagers along the Sesan and Srepok Rivers upstream of LS2, as well as 87 villages along tributaries of the two rivers, losing access to migratory fish. The same research also found that over 22,000 villagers living downstream from LS2 would be negatively impacted as a result of changes in the river hydrology and water quality (Baird 2009). It has been

[13]This region is known as the "3S region" by taking the initial "S" of the names of the three rivers in English. They are international rivers that flow into northeastern Cambodia from their sources in the plateaus of central Vietnam and Laos. The dominant residents in this area are not ethnic Khmer, but indigenous and ethnic minority groups from Cambodia, such as Punong, Brao, and Lao ethnicities (3 SPN and Mekong Watch 2015).

[14]Information from a local resident, September 8–12, 2015.

estimated that the project could produce a 9.3% drop in fish stocks across the whole of the lower Mekong basin (Ziv et al. 2012).

LS2 is a build–operate–transfer (BOT) project and construction is expected to take 5 years. After 40 years of operation, its ownership will be transferred to the government of Cambodia. According to the government's announcement, generated power will be sold to the Electricite du Cambodge (EDC), but it may still be exported to Vietnam. Project developers are the Hydropower Lower Sesan 2 Company, a joint venture between Cambodia's Royal Group (39%) and China's Hydrolancang International Energy (51%), and EVN International Joint Stock Company (EVNI), a subsidiary of the Electricity of Vietnam (EVN) (10%) (3SPN and Mekong Watch 2015).

Given the importance of this project for the economy of a developing country like Cambodia, affected communities in Ratanakiri Province and Stung Treng Province, however, have not been adequately consulted. Basic information on LS2 itself is rare and difficult to obtain, furthermore, information or assessments on its negative impact are not publicly accessible. Some parts of the resettlement plan or compensation programs are publically open, however, they are not understandable for affected people. Villagers living downstream of LS2 have not been consulted in regards to resettlement and compensation issues during the decision-making process. Affected people who were forced to relocate were only informed after relevant plans or policies were approved and they were not given any opportunities for discussion (3SPN and Mekong Watch 2015).

The so-called "international standard" provides that local people should participate in the first stage of the project. This guideline is fully acknowledged by the World Bank,[15] the Asian Development Bank,[16] and Japan International Cooperation Authority (JICA)[17] to protect people's rights informing them of how they would be affected in advance. In this case, however, affected communities in the upstream area, mainly in Ratanakiri Province, did not even know that construction had commenced.[18] Furthermore, there is also much criticism whether such a project will really protect the environment and the livelihood of people as it claims to.[19]

In November 2012 and February 2013, local authorities held a meeting in villages inside the reservoir area. According to villagers who attended, the authorities

[15]World Bank Safeguard Policies can be found online at the site's homepage. http://web.worldbank.org/WBSITE/EXTERNAL/PROJECTS/EXTPOLICIES/EXTSAFEPOL/0,,menuPK:584441~pagePK:64168427~piPK:64168435~theSitePK:584435,00.html (accessed on August 22, 2016)

[16]ADB Safeguard Policies: http://www.adb.org/site/safeguards/main (accessed on August 22, 2016).

[17]JICA Environmental Guideline: http://www.jica.go.jp/english/our_work/social_environmental/guideline/index.html (accessed on 22 August, 2016).

[18]Interview with villagers in Ratanakiri Province, September 8–12, 2015.

[19]International Rivers; https://www.internationalrivers.org/campaigns/lower-sesan-2-dam, Southeast Asia Globe; http://sea-globe.com/lower-sesan-2-dam-communities-threatened-southeast-asia-globe/, EarthRights International; https://www.earthrights.org/media/lower-sesan-ii-dam-developers-urged-ensure-accountability-severe-impacts.

proposed new resettlement site options, including a place along the Sekong River and explained that compensation would include land for houses and agriculture. However, villagers complained that the information was still ambiguous and untrustworthy because detailed information and written documents were not provided (3SPN and Mekong Watch 2015).

In R Village, one of the affected villages located along the Srepok River, public tension about the project is high. Most of its inhabitants are of the indigenous Punong ethnic minority. During a field visit in February 2015, many villagers were voicing their opposition to the dam's construction, and so far there has been no agreement between developers and residents. However, it came to light that an asset survey of the project had commenced and that villagers were anxious about the current situation expressing fear of both the survey and dam construction. One villager in his 20s explained the measurement process in detail:

> When the survey group came, they did not explain or show anything to us. They just asked the house owners or their families the area of their house and how many mango trees or coconut trees they had, and recorded the data. They also asked whether they wanted a concrete or wooden house. Some answered they wanted money to construct houses by themselves. They sprayed on the house and trees for which they will pay compensation. After that, they asked both husband and wife to place their thumbprints. Then they took photos of family and their house.

He continued to share why families had their houses measured.

> Some families who had their houses measured understood the meaning of the measurement, but some did not understand clearly. They were afraid of being flooded by the dam, so they just followed those who allowed their houses to be measured. Some said they did not want to move out but they had no choice. They did not want to have their houses measured, but they were persuaded by the measuring group, who were accompanied by the police (Mekong Watch 2015).

Another villager in her 30s shared her case.

> "People complain to me about why I oppose LS2. They say I shouldn't because my family member is a government worker. We were threatened we won't be receiving a salary if we oppose LS2. Even some relatives blame us and say we are stupid for not accepting money. This compensation issue is breaking up relationships among relatives" (Mekong Watch 2015).
>
> As of June 2016, there are around 200 families who have not agreed to the government resettlement plan.[20]

One major concern of the affected communities is fishing, an issue that has, unfortunately, been ignored in the government plan and received little attention. For affected communities, the fishing issue cannot be left out of discussions as their livelihood depends on it. Mr. Atith (a pseudonym), resident of Koun Mom District, Ratanakiri Province, speaking on the bank of the Srepok River, said that

[20]Information in the meeting, June 2016.

> Even if we put out the gill nets, we will get no fish. Until now, if we set nets in the morning, fish would be caught in them by the evening. I have no idea where the fish have gone. Where I used to catch five kg of fish each time, it is now half that. Several years ago, we were able to catch a lot more fish. This year, people in other villages and downstream are seeing the same problem…. Water is being held back by the dam construction and the fish may have gone to another river.[21]

On a separate occasion, Mr. Chamroeun (pseudonym) of Lumphat District also confirmed what Mr. Atith said about the situation in his village: the Mekong migratory fish that swim upstream around May have not come up in the Srepok River. "Until a few years ago, I was just about able to catch enough fish to feed my family. But our life has become very hard since May (because there are no fish to catch)."[22]

An elderly man who lives in an ethnic Lao village in the Andoung Meas district admitted that people in his community have been suffering from reduced numbers of fish, deteriorating water quality and flooding for many years due to the dams built upstream in Vietnam. He regrets the dam construction without prior environmental assessment when he said that "As things stand now, we do not want them to build the dam. The dam will cause the fish to disappear and will destroy nature. We want them to make a full account of the impacts before the dam is built, and we do not want them to construct until that is done."[23] He has a genuine grievance that the dam construction has ignored the community's voices and without prior knowledge and involvement from the affected community.

A number of accounts by villagers in Ratanakiri Province confirm that many families in the area have not even been informed about the construction of LS2.[24] When they heard about it, they collectively said that both the government and the company should explain the project to them directly. They were seething with anger because they felt that no one cares about their opinions. It is thought that around 78,000 upstream residents, mostly in Ratanakiri Province, will be the most affected by the construction of LS2 (Baird 2009).

Companies involved in the LS2 construction are said to be considering the construction of a fish pass (Gatke 2013), but while the construction of the dam has begun, it is unclear whether a fish pass is being incorporated or not. Far worse, the final design and size of the dam itself is unknown because even such basic information has not being disclosed to the public.

Conclusion: different projects, the same story

This chapter has shown the importance of local small-scale fisheries for people's livelihoods and their culture in the Mekong river area. While there has already been a great amount of discussion about the negative impacts of dams on people's livelihood and way of life along the river (Baran 2010; Baird 2009; Roberts 1993,

---

[21]Interview, on villagers in Ratanakiri Province, September 8, 2015.

[22]Interview, on villagers in Ratanakiri Province, September 9, 2015.

[23]Interview, on villagers in Ratanakiri Province, September 10, 2015

[24]Based on author's fieldwork in September, 2015.

2001; Ubon Ratchathani University 2002), the issue of fisheries is too often ignored and not touched upon in terms of fish migration patterns. As this chapter has described from the Pak Mun Dam case, there are other important concerns for affected communities. These include the irretrievable loss of local knowledge and fishing skills, forced changes to their occupation, and threats to their main income sources.

While a fish pass has been introduced at the Pak Mun Dam (and will be introduced in Lower Sesan 2 Dam) as a solution to mitigate negative effects on fishery, there is no single study in the Mekong River basin that confirms if fish can really reproduce after crossing a dam through a fish pass. On the other hand, after a dam changes the river ecology as the Pak Mun Dam does, fishermen's knowledge is made redundant as there are no fish to catch. Communities depend on fish resources and inevitably face a whole host of challenges where dams are constructed. Fishing skills and knowledge of the ecology of rivers are lost.

Should the Pak Mun Dam case provide a cautionary tale on dam construction in the Mekong river region, there is little evidence that the Lower Sesan 2 Dam has taken note. With Pak Mun Dam, villagers in a remote area of Cambodia have had few chances to access information through television or newspapers regarding the Lower Sesan 2 Dam. There is insufficient information before dam construction, and in some cases, the withholding of it. When a national project comes to the village, authorities (both national and local) only explain the benefits to affected communities (job possibilities, new homes, and improved road access). The story for the affected people of the, i.e. Pak Mun Dam and the, i.e. the Lower Sesan 2 Dam is not so different.

The residents of the Sesan and Srepok River basin in Ratanakiri Province, who are one of the stakeholders in the dam project, are completely uninformed about any measures that might be taken to alleviate impacts. The one public hearing on the dam, which resulted from pressure by a citizens group, was held in Stung Treng Province, where the dam is being constructed. Pak Mun dam's affected people rose up after construction had finished. In 2015, there are around 2500 families still demanding the opening of the gate and damage compensation for the operation period.[25] The fishing communities of Ratanakiri Province may witness and experience similar tension and struggle.

Even though a number of international aid agencies have provided safeguard policies or social environmental guidelines, dam projects in the Mekong Region keep disrupting the environment and livelihood of locals. Project owners and funders may change, but local communities who give all too familiar testimonies bear witness alone. As such, further serious reconsideration of how academia can engage in these issues is needed to find the best way to protect people's livelihood through empirical research that draws out local knowledge and its interaction with environmental conditions along the Mekong River.

[25]Interview with a supporter of the Pak Mun affected people in October 28, 2015.

Acknowledgements At the field research, Dr. Kanokwan Manorom, Ubon Ratchathani University and Ms. Sompharn Khundee, Supporter of Assembly of the Poor helped me to conduct fieldwork in Thailand. Dr. Akihisa Iwata, Professor at the Graduate School of Asian and African Area Studies (ASAFAS), Kyoto University, provided valuable advice for collecting data. 3S Rivers Protection Network (3SPN) helped me to interview villagers in Ratanakiri Province, Cambodia. Colleagues of the Mekong Watch provided invaluable support to continue working on this issue. This survey would not have been possible without those who kindly allowed me to interview them, and I thank them all. I also would like to express my appreciation to Ms. Wanida Thantiwithayaphitak. Her devotion to those villagers who have suffered from the Pak Mun Dam has inspired me to keep researching this issue. Field research was supported by the Keidanren Nature Conservation Fund and two ASAFAS programs, Initiatives for Attractive Education in Graduate Schools and the 21st Century COE program: Aiming for COE of Integrated Area Studies. At Ubon Ratchathani, The Mekong Sub-region Social Research Center (MSSRC) assisted to provide me with formal research status to conduct field surveys in Thailand.

References

3S Rivers Protection Network (3SPN) and Mekong Watch. (2015). *Fact Sheet: Lower Sesan 2 Hydropower Project*. Northeastern Cambodia: 3SPN and Mekong Watch.

Baird, I. G. (2009). *Best practices in compensation and resettlement for large dams: The case of the planned lower Sesan 2 hydropower project in northeastern Cambodia*. The River Coalition in Cambodia, Phnom Penh: Cambodia. https://www.academia.edu/1049246/Best_Practices_in_Compensation_and_Resettlement_for_Large_Dams_The_Case_of_the_Planned_Lower_Sesan_2_Hydropower_Project_in_Northeastern_Cambodia. Accessed: 22. Aug. 2016.

Baran, E. (2010). *Mekong fisheries and mainstream dams. Fisheries sections in: ICEM 2010. Mekong River Commission Strategic Environmental Assessment of Hydropower on the Mekong mainstream*. Hanoi & VietNam: International Centre for Environmental Management.

Baran, E., Jantunen, T., & Chiew K. C. (2007). *Value of inland fisheries in the Mekong river basin*. Phnom Penh: WorldFish Center. http://pubs.iclarm.net/resource_centre/WF_895.pdf. Accessed: 22. Aug. 2016.

Baran, E., et al. (2013). *Fish And fisheries In the Sekong, Sesan And Srepok basin (3S Rivers, Mekong watershed), with special reference to the Sesan river*. Phnom Penh: World Fish Center.

Claridge, G. et al. (1997). *Community fisheries in Lao PDR: A survey of techniques and issues*. Vientiane: The World Conservation Union (IUCN).

Foran, T., & Manorom, K. (2009). Pak Mun Dam: Perpetually contested? In F. Molle, T. Foran, & M. Kakonen (eds.), *Contested waterscapes in the Mekong river: Hydropower, livelihoods and governance* (pp. 55–80). London: Earthscan.

Gatke, P. et al. (2013). *Fish passage opportunities for the Lower Sesan 2 Dam in Cambodia – lessons from South America*. Project report. Hanoi: International Centre for Environmental Management (ICEM). https://wle-mekong.cgiar.org/download/mk3-optimizing-cascades/MK3-Fish-passage-opportunities.pdf. Accessed: 22. Aug. 2016.

Kanokwan, M. (2004). *Atalak chang sangkhom chao pramong mae nam mun ton lang: mun phitcarana cak rabop niweat* [Identities of fisheries in the Lower Mun River: Focus on its ecology]. In W. Prachayakan (ed.), *Watakam atalak* (pp. 71–115). Sirinthorn anthropology center (in Thai).

Kiguchi, Y. (2006). The Impacts of the Pak Mun Dam's construction on fishing and livelihoods in the Lower Mun river region of Northeast Thailand. Masters thesis, Graduate school of Asian and African Area Studies (ASAFAS), Kyoto University (in Japanese).

Kiguchi, Y. (2014). The change in the usage of *toum yai* fishing gear in north eastern Thailand. In Mekong sub-region social research center (ed.), *Changing way of life of ethnicities in the Mekong region* (pp. 105–125). Ubon Ratchathani: Faculty of Liberal Arts, Ubon Ratchathani University.

Mekong River Commission. (2002). *Fish migrations of the Lower Mekong river basin: implications for development, planning and environmental management.* Phnom Penh: Mekong River Commission.

Mekong River Commission. (2007). *Consumption and the yield of fish and other aquatic animal from the Lower Mekong Basin. (MRC Technical Paper No.16).* Phnom Penh: Mekong River Commission.

Mekong River Commission. (2010). *State of the Basin Report*. Phnom Penh: Mekong.

Mekong Watch. (2004). *A River, its fish and its people: Local knowledge of the national environment at the mouth of the Mun River*. Tokyo: Mekong Watch.

Mekong Watch. (2015). *Interview at affected village in February 2015 regarding asset survey held on Lower Sesan 2 Dam (Field Note).* Tokyo: Mekong Watch.

Research Committee of the Villagers of Rasi Salai. (2004). *Rasi Salai: Traditional knowledge, rights and ways of the Mun river's flood forest.* Three-Province People's Organization Network for protecting wetlands in the Mun River (in Thai).

Roberts, T. R. (1993). Just another dammed river? Negative impact of Pak Mun Dam on fishes of the Mekong basin. *Natural History of Bulletin of the Siam Society,* 4, 105–133.

Roberts, T. R. (2001). On the river of no returns: Thailand's Pak Mun Dam and its fish ladder. *Natural History of Bulletin of the Siam Society, 49,* 189–230.

TEAM Consulting Engineers. (1982). Environmental and ecological impact assessment of Pak Mun Project, Vol. 1, Summary Report.

The Assembly of the Poor (AOP) and Southeast Asia rivers network (SEARIN). (2002a). *Mae Mun: Kaan klap ma khong khon ha pla [The river Mun: Return of the fishing people].* Chiang Mai: Southeast Asia Rivers Network. (in Thai).

The Assembly of the Poor (AOP) and Southeast Asia rivers network (SEARIN). (2002b). *Mae Mun: Kan klap ma khong khon ha pla: bot sarup lae khwaam ruu ruang plaa khong khon paak muun* [The river Mun: Return of the fishing people: Summary of local knowledge of fish]. Chiang Mai: Southeast Asia Rivers Network. (in Thai).

Thompson, C. (2008). *Wild Mekong Report.* WWF Greater Mekong Programme. http://www.wwf.be/_media/greater_mekong_species_report_web_ready_version_nov_14_2011_1_678824.pdf. Accessed: 22. Aug. 2016.

Ubon Ratchatani University. (2002). *Khrong kan sueksa naew tang fuen fu rabop niwet withi chiwit lae chumchon thi dai rap phon krathop chak kan sang khuean pak mun* [Project to study approaches to restoration of the ecology, livelihood, and communities receiving impacts from construction of the Pak Mun Dam]. Ubon Rachatani: Ubon Ratchatani University (in Thai).

World Commission on Dams (WCD). (2000). *WCD case study: Pak Mun Dam Mekong River Basin, Thailand.* World Commission on Dams.

Ziv, G. et al. (2012). Trading-off fish biodiversity, food security, and hydropower in the Mekong River Basin. Proceedings of the National Academy of Science 109.15: -5614. http://www.pnas.org/content/109/15/5609.full. Accessed: 22 Aug. 2016.

Chapter 11
Livelihood Activities of Swiddeners Under the Transition of Swidden Agriculture: A Case Study in a Khmu Village, Northern Laos

Nyein Chan, Lamphoune Xayvongsa and Shinya Takeda

Abstract The proportional extent of swidden cultivation in Laos is greater than in any other Southeast Asian country. Yet since 1975, the government of Laos has made attempts to eradicate, or at least reduce, swidden cultivation through a village relocation policy predicated upon security concerns, through a "shifting cultivation" eradication policy, and a land and forest allocation program. Recently, the Land Allocation Programme (LAP), known as the "Three-Plot Policy," was introduced across the country. This chapter presents a case study from the Khmu area, northern Laos, where land available for swidden culture was limited by LAP implemented in 2011. The chapter shows how this policy created difficulties for villagers to pursue traditional swidden practice. It presents the crucial role swidden cultivation plays in the livelihoods of the Khmu swidden cultivators and highlights a notable shift from traditional to semicommercialized practices. This chapter clarifies the possible impacts of changes in swidden and argues that any future planning of REDD+—as the government has attempted to promote it in swidden fallows, requires detailed observation of the role and function of swidden cultivation for livelihood purposes.

Keywords REDD+ · Khmu swiddeners · Land and forest allocation programme Semicommercialization · Livelihood

N. Chan (✉)
Department of Pollution Control and Waste Management, University of Forestry and Environmental Science, Yezin, Nay Pi Taw, Myanmar
e-mail: nchan08@gmail.com

L. Xayvongsa
Faculty of Forest Science, National University of Laos, Laos, Lao People's Democratic Republic

S. Takeda
Graduate School of Asian and African Studies (ASAFAS), Kyoto University, Kyoto, Japan

M. Lopez and J. Suryomenggolo (eds.), *Environmental Resources Use and Challenges in Contemporary Southeast Asia*, Asia in Transition 7,
https://doi.org/10.1007/978-981-10-8881-0_11

Introduction

Recent scholarly attention has focused on swidden agriculture as a new mechanism for Reducing Emissions from Deforestation and forest Degradation (REDD +), while taking into consideration the needs of local communities and ensuring long-term sustainability and equitable benefit sharing. Such a new mechanism may present either a challenge or an opportunity for low-income countries like Laos[1]. Within such a context, it is essential to understand the livelihood situation of swiddeners, whose traditional practice has undergone a transition over the past three decades.

Swidden, even the system itself, is a diverse, complex and dynamic form of land use. This has made it difficult for researchers to provide a clear definition, and measure and quantify (Schmidt-Vogt et al. 2009). Governments (including colonial, and postcolonial) have tended to see swidden agriculture as a primitive anachronism that destroys forests and as such has been the target of "eradication" to be replaced as soon as possible with permanent fields (rubber plantations, maize farming, etc.) through the logic of scientific resource management (Ellen 2012). It is in this context that scholars have critically noted that swidden agriculture in the global tropics, especially in Southeast Asia, have been transformed, or wholly replaced by other forms of land use. This has meant that in some regions it has almost disappeared due to political, economical, and sociocultural factors (Padoch et al. 2007; Fox et al. 2009; Schmidt-Vogt et al. 2009; van Vliet et al. 2012; Ziegler et al. 2009).

As in other tropical countries, swidden agriculture in Laos is also being rapidly replaced and transformed into agricultural practices that prioritize other land uses (Higashi 2015; Kameda 2015). The nation has seen a series of government policies introduced with the purpose of promoting conservation, the sustainable use of forest resources, and poverty alleviation (Souvanthong 1995; Hansen 1997; Fujisaki 2012; Higashi 2015). Based on various internal documents from the Ministry of Agriculture and Forestry of Laos (Kitamura 2004), the extent of swidden agriculture and the number of households engaged in it decreased from 176,605 ha and 186,265 households in 1996 to 118,900 ha and 174,036 households in 2000. This change makes it clear that the government has attempted to stabilize swiddens in the country (Souvanthong 1995; Hansen 1997; Fujisaki 2012; Higashi 2015). It is important to note that in the 1990s, the Land and Forest Allocation (LFA) program was initiated across the country promoting an "eradication policy" for swidden agriculture. As a result of this, efforts to stabilize the practice of swidden agriculture have had a range of impacts on local indigenous communities and on the environment (Higashi 2015).

[1]Laos covers an area of approximately 230,000 km^2 and is sharing borders with Myanmar and China in the northwest, Vietnam in the east, Cambodia in the south, and Thailand in the west.

Among the country's indigenous communities, the Khmu has been particularly affected by various programs (Higashi 2015). However, few studies have assessed the livelihood of swidden cultivators in northern Laos after the introduction of LFA programme. A number of scholars have produced some interesting and detailed reports about the situation. Higashi (2015) studied land and forest management by swidden cultivators in community-based watershed forests in Oudomxay Province, northern Laos. In addition, Anoulom (2008) assessed livelihood, but only related to non-timber forest products (NTFPs). Kameda (2015) identified fallow periods and weed problems as a major factor influencing the changes taking place in swidden agriculture. This chapter discusses the livelihood situation of Khmu swiddeners, and evaluates the potential impacts of the LFA programme on their livelihood activities. This research agrees with previous studies and notes the deep impact of the program on the livelihood of these indigenous communities, especially that it became difficult for villagers to pursue traditional swidden practices. However, based on fieldwork in the area, this chapter shows the fundamental role of swidden cultivation for the livelihood of the Khmu, and how the shift from traditional practices to a semicommercialized level is accepted and practiced by the community as a whole.

In addition, this chapter highlights the importance of community participation in any government planning, as in any other low-income countries elsewhere, especially for policies regarding the roles and functions of swidden cultivation. The government of Laos has recently promoted the international mechanism of Reducing Emissions from Deforestation and forest Degradation (REDD+) and enhancing forest carbon stocks (Mertz et al. 2009a; Hett et al. 2012), while simultaneously promoting rural development (Anoulom 2008; Higashi 2015), especially in areas of swidden cultivation. This case study of the Khmu community clarifies the possible impacts of changes in swidden cultivation in Laos, and also contributes to an integrated understanding regarding the socioeconomic condition of swiddeners to better planning at local and/or regional level.

Change of Swidden Agriculture in Southeast Asia

A number of important studies have attempted to estimate the size of swidden agriculture in Southeast Asia, and estimates suggest that nearly 50% of land area in the region is under swidden (Spencer 1966; Schmidt-Vogt et al. 2009; Ellen 2012). Of these, Schmidt-Vogt et al. (2009) and Mertz et al. (2009b) have reviewed and assessed the extent of swidden and the number of swiddeners in the region. Swidden areas are estimated to account for 1 million ha in Thailand, 2.0–6.5 million ha in Laos, 0.349 million ha in Cambodia, one-fifth of the total land area in Vietnam, 0.29–10.18 million ha in Myanmar, 2.3 and 1.4 million ha in Sarawak and

Sabah of Malaysia (based on 1980s data), and 11.4 million ha in Indonesia (including East and West Timor) (See Schmidt-Vogt et al. 2009 for detailed discussion). Estimates based on demographic data suggest there are around 14–34 million people who are swidden cultivators (See Mertz et al. 2009b). These offer a partial assessment based on recently available updated data of swidden and its demography in the region.

Schmidt-Vogt et al. (2009) reported two patterns of change in swidden: a shortening fallow length and the replacement of swidden systems with other types of land use. A review of 55 swidden systems (Schmidt-Vogt et al. 2009), showed that 40 cases are of shortening fallow length, whereas the remaining revealed a constant fallow length. This review also notes that 91 cases of swidden systems were replaced with other forms of agriculture and other types of livelihood systems.

Fox et al. (2009), based on an analysis of case studies, reported driving factors of swidden changes. These are the classification of swiddeners as ethnic minorities within nation-states (with a negative assessment); the division of landscape into forest and permanent agriculture in land policies and laws (i.e., the nonappearance of swidden even as agroforestry systems on maps appear as agriculture and forests); the expansion of forest departments and the rise of conservation in forested area, where swidden is practiced; resettlement (immigration and emigration); the privatization and commoditization of land and land-based production, which contrasts to swiddeners' communal and customary forms of land tenure; and the promotion of market-driven and industrial agriculture. Meanwhile, Cramb et al. (2009) added sociocultural causes to the decline of swiddens such as the rise of modern education systems, conversion to mainstream religions, and exposure to mass media and changing ideologies.

Despite methodological differences and perspectives, scholars agree that the change in swidden land use will affect the livelihood of swidden cultivators. Cramb et al. (2009) reviewed food security and other resource-based livelihood activities of swiddeners in Southeast Asia (six case studies in the Philippines, Indonesia, Vietnam, and Thailand), reflecting various changes in livelihood of different ethnic swiddeners affected by swidden transformation. In Palawan, Philippines, the combination of agroforestry and paddy rice fields in suitably flat lands for wealthier swiddeners seems to have been successful[2], but the poor were likely to be locked into swidden production due to limited access to productive resources and political networks (Cramb et al. 2009; Dressler and Turner 2008; Dressler 2006). In East and West Kalimantan, Indonesia, supportive measures such as wage labor opportunities in oil palm plantation, education and medical services were provided to local people, whereas access to swidden was permitted for subsistence (Cramb et al. 2009; Colfer 2008; Peluso 2005; Colfer et al. 1997; Peluso 1996; Colfer and Dudley 1993). Furthermore, rotational swidden was practiced in communal land as

[2]Cramb et al. (2009, p. 332–333) notes that the conservation initiatives by NGOs and government officials have made a preferential change of the wealthy Tagbanua farmers to intensify swidden agriculture through ecological ethnic groups.

a safety net, while there was shift to rubber agroforestry in Jambi, Indonesia (Cramb et al. 2009; Cramb 1993). In the Central Highlands of Vietnam, development interventions—in which migration and agricultural intensification were sponsored by the State—was not successful due to decreasing participation willingness by local swiddeners in the long run (Cramb et al. 2009; Salemink 2003). Finally, the role of local institutions in the multiple responses[3] to the regulation of swidden in Northern Thailand is also a factor (Cramb et al. 2009; Laungaramsri 2005; Schmidt-Vogt 1998; Grandstaff 1980).

What is clear is that the transformation of swidden in Southeast Asia and the change in swidden supporting livelihood is not a uniform process in space or time (Cramb et al. 2009; Fox et al. 2009; Schmidt-Vogt et al. 2009). As a result of the diverse processes and forces at play, various impacts on livelihood will vary. In some cases, the increasing difference within and between communities could leave some groups marginalized and worse off. Under certain situations, commercial agriculture (e.g., the introduction of cash crops) could help improved household livelihoods, but market instability might increase vulnerability (Cramb et al. 2009). In some cases, swidden still plays an important role in local communities livelihood strategies by providing resilience in the face of serious changes (e.g. El Niño event in East Kalimantan, Indonesia in 1972 and 1983, following a rat infestation). Therefore, understanding the socioeconomic condition of swidden cultivators is essential in planning, implementing, monitoring, and evaluating both development and conservation program in swidden lands.

Swidden Agriculture in Laos

As a landlocked country in mainland Southeast Asia, the landscape in Laos comprises mountainous areas with steep slopes and rugged terrain, and lowland alluvial plains (Hett et al. 2011, 2012; Hurni et al. 2013; Heinimann et al. 2013). Almost 80% of the land is hilly and mountainous, and about 40% of the total population lives in the mountainous areas (Souvanthong 1995; Fujisaki 2012).

Estimates by Mertz et al. (2009a, b), state that around 800,000–1,000,000 people (approximately 20% of the rural population) in Laos still engage in swidden agriculture. In particular, the landscape of northern Laos has been dominated by swidden fields and fallow forest mosaics (Fox et al. 2009; Spencer 1966). Using a landscape mosaic approach[4], Messerli et al. (2009) reported that the extent of

[3]Responses of local institutions to the state's pejorative perception of swidden agriculture has been varied. Some examples of local responses are political movement by NGOs and local people, 99-day demonstration by members of Northern Farmer's Network, and so on (Cramb et al. 2009: 19–20).

[4]Voluminous land use literature covers several case studies dealing with human—environment interactions only at local levels. Only few studies were observed to tackle the issue of linking land cover change to underlying processes at higher spatial scale. Hotspot approach and metadata

swidden in Laos was 6,500,000 ha (28.2% of the total area). Based on this fact, the proportional extent of swidden is probably greater in Laos than in any other country in Southeast Asia (Schmidt-Vogt et al. 2009; Ellen 2012).

Restriction of swidden agriculture is one of the current legal forest preservation activities in Laos (Government of Laos [GOL] 2007). Attempts have been made by the government of Laos to "eradicate," or at least reduce, swidden cultivation since 1975 through a village relocation policy due to security concerns, and through a "shifting cultivation eradication policy" and a land and forest allocation program (Kenney-Lazar no date; Higashi 2015; Kameda 2015). Recently, due to the introduction of the LFA program, also known as the "Three-Plot Policy," the mean fallow period for swidden agriculture in northern Laos decreased from around 20 years in the 1970s to around 5 years in the 1990s (Roder 1997). For the case study, area presented in this chapter (Khmu area, northern Laos), the LFA program, in which the area available to swidden cultivation has been limited, was implemented in 2011, subsequently making it difficult for the villagers to pursue their traditional swidden practice.

Livelihood Activities of Swiddeners in the Transition of Swidden Agriculture

This case study was carried out in S village in Viengkham District, Luang Prabang Province, Northern Laos (Fig. 11.1). The village lies about 204 km from the provincial capital of Luang Prabang. It contains a total of 65 households and had a population of 381 as of 2014. All villagers belong to the Khmu ethnic group.

The village was established in 1975 by Khmu people from Pakxeng. In 1977, they temporarily moved to Viengkham for one year, and subsequently resettled in the village. The total land area belonging to the village is 1961.29 ha. Within the village boundaries, the mainstream, Huay Sangiao, flows from the southeastern to the northwestern area. Based on the designations of the LFA program initiated in the village in 2011, there is a livestock breeding zone (385.57 ha) assigned on the north side of the stream, and forests are protected for watershed management (131.66 ha) along its small tributaries. The forests around the settlement area are managed for utilization purposes (141.39 ha). Conservation forests occupy about 166.28 ha in the southwestern part of the village. Most of the permanent farms (108.09 ha) lie along the main road. In addition, there is a rubber plantation of 131.88 ha, and 789.23 ha is utilized for swidden cultivation in the remaining part of the village (Provincial Agriculture and Forestry Office [PAFO] 2011) (Fig. 11.2).

---

analysis approach are noteworthy. Landscape mosaic approach reveals the information about the underlying social and environmental processes and hence, humanenvironmental interactions. It should allow not only the integration of land cover inventories with land use processes over larger areas but it should also offer the potential of contributing to the generalization of knowledge, aggregation, and scaling (Messerli et al. 2009 pp. 292–293).

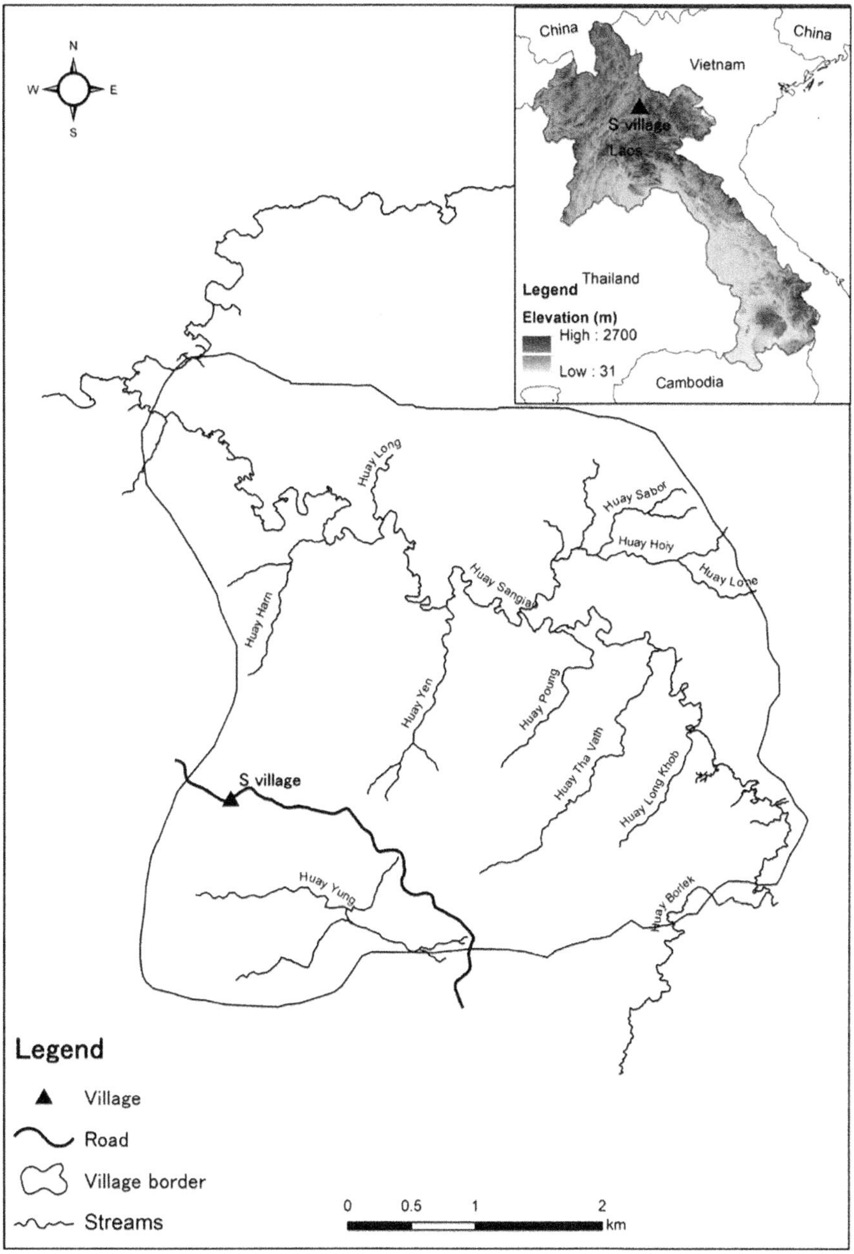

Fig. 11.1 Location of case study in Luang Prabang province, Northern Laos

Fig. 11.2 Land use map of S village

Due to the village's geography, there are very limited available flat lands for paddy farming. The major livelihood activity in the village is swidden cultivation, with an average rotation of 4–6 years and average plot size of 1.71 ha (Takeda 2007), which were mainly observed between 550 m and 875 m above sea level. Cultivation begins with site selection during early February and continues with the practice of slashing and burning around April, planting around May, and finally, harvesting around November. The major crop is upland rice. In addition, several other crops including maize, chili, sweet potato, taro, pumpkin, gourd, and melon are grown for home consumption. Recently, the production of maize and black rice as cash crops have increased. After harvesting, the land is left fallow. Various kinds of NTFPs, such as bamboo shoots, mushrooms, and small animals for self-consumption and for sale are collected from the forest fallows. Livestock (such as water buffalo, cattle, pig, goat, and poultry) is also one source of cash income for members of the community.

The climate in the region is tropical monsoonal, with the rainy period occurring between April to October (Fitriana 2008). The region lies in the hill evergreen forests, with elevations ranging from 500 to 900 m above sea level. At the Luang Prabang meteorological station, the average annual precipitation over the past 5 years was about 1424 mm, and the mean monthly temperature about 25 °C

(Weather Underground database). According to the FAO/United Nations Educational, Scientific and Cultural Organization (FAO/UNESCO 1979), the soil type in this region belongs to Orthic Acrisols.

The number of households engaged in swidden in this village decreased from 84 households in 2005 to 56 households in 2011 (Takeda 2013). The extent of swidden also decreased 143.65 ha in 2005 to 82.63 ha in 2011, with the exception of 139.78 ha and 112.70 ha in 2008 and 2009, respectively (Takeda ibid), probably due to high price of maize in the world market (Table 11.1; Fig. 11.3). The average fallow period in this village reduced from 5 years in 2005 to 3 years in 2011 (Takeda ibid).

Using semi-structured questionnaires, all the households (65 in total) were interviewed in September 2014 about their land use and agricultural crop production. Crop harvesting for the year 2014 had not yet been carried out at the time of data collection; therefore, the crop production for 2013 was collected as recalled data. Based on interview data, the average production of agricultural crops was analyzed. For crop production, agricultural crops such as rice, maize, and cassava, and fallow crops such as broom grass (*Thysanolaena latifolia*) were categorized because the swidden farmers earned cash mainly from these crops.

According to the interview surveys, 56 respondent households (86% of total households) were engaged in swidden practice in 2013 (Table 11.2). As cash crops, 45 respondent households (69%), 7 households (11%), 3 households (5%), and 1 household (2%) produced maize, black rice, cassava, and *mak kha* (*Alpinia galangal*), respectively. Cash crops were produced not only in agricultural plots, but also in fallow plots. Four households (6%) produced broom grass from their fallow plots as a cash crop. The average production of each type of agricultural and fallow crop is outlined in Table 11.2. However, cash crop production is variable and depends mainly on market conditions.

Upland rice was produced mainly for subsistence. In 2013, the average rice production and consumption per household were 1816 and 1095 kg, respectively. According to the interviews, 31 households (48%) were able to produce enough rice for home consumption; however, 34 households (52%) were unable to meet their consumption demands.

Table 11.1 Maize production in Laos (2005–2014)

Year	2005	2006	2007	2008	2009	2010	2011	2012	2013	2014
Area harvested (0.000 ha)	86.0	113.8	157.7	229.2	207.6	212.7	212.1	196.8	212.0	243.4
Yield (ton/ha)	43.3	39.5	43.6	48.3	54.6	48.0	51.7	57.2	57.3	58.0
Production (mil. ton)	0.4	0.4	0.7	1.1	1.1	1.0	1.1	1.1	1.2	1.4
Seed production (0.000 ton)	2.3	3.2	4.6	4.2	4.3	4.2	3.9	4.4	4.4	–

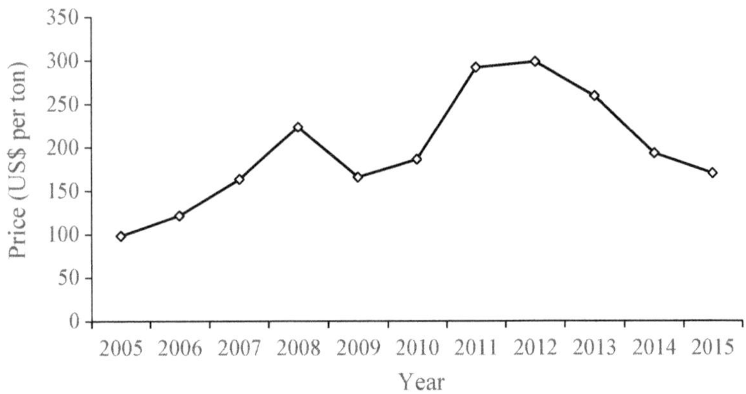

Fig. 11.3 Price trend for maize during 2005–2015 (*Note* Data was extracted from www.imf.org, based on Maize (corn), U.S. No. 2 Yellow, FOB, Gulf of Mexico)

Table 11.2 Average production of agricultural crops in S village, northern Laos in 2013

Crops	N (HHS)	Production (kg per HH)
Agricultural crops Rice	56	1832.50 ± 147.742
Maize	45	2263.1 ± 258.68
Black rice	7	765.7 ± 198.35
Cassava	3	2433.3 ± 1105.04
Mak kha[a] (*Alpinia galangal*[b])	1	150.0
Fallow crop		
Broom grass	4	122.5 ± 26.58

Note [a]Local name for one agricultural product; [b]Scientific name for *Mak kha*

The major income-generating sources include agricultural produce, non-timber forest products (NTFPs), animal husbandry, and others. Though the upland rice is their major subsistence crop, maize production increased in this village. Interviews clarified that 45 households produced the average maize production of >2000 kg in 2013. Furthermore, black rice has become a promising cash crop in the village because it can be sold in Luang Prabang, a tourist destination. In 2013, seven households were producing black rice. *Mak kha* was planted in the swidden fields for cash income, with an average production of 150 kg was reported in 2013. Therefore, in this village, traditional swidden practice for subsistence purposes has gradually transformed into semicommercialized farming.

The maize production in Laos during 2005–2014 was described in Table 11.1, referring FAOSTAT. The area harvested and yield of maize markedly increased since 2008, based on commodity prices (Fig. 11.3). The global trend of maize price seems to largely affect even a small village like S village.

NTFPs played a significant role in the livelihood activities of swiddeners. In 2013, bamboo shoots were collected as a cash-generating crop by 44 households, broom grass by 34, *peuak-meuak* (*Boehmeria* sp.) by 33, paper mulberry (*Broussonetia papyrifera*) by 27, *pong-peng* (*Flickingeria* spp.) by 10 households, rattan shoots by 3, and bamboo insects by 1 household (Fig. 11.3).

In addition, animal husbandry was actively carried out by the swiddeners in this village with the aid of local government and non-governmental organizations (NGOs) and international non-governmental organizations (INGOs). Animals traditionally reared and promoted by the development program include cattle, water buffalo, goat, pigs, and poultry. Table 11.3 details the number of animals reared in the village. According to farmers, most cash is earned by selling domesticated animals only in difficult times.

Other income sources include family remittances, government support, and home businesses. Most households (31% of the total) received income from the remittances of family members who were working outside the area. Also, land rental has recently been introduced in the village as an income-generating activity, although only one household was earning an income in this way.

The results show that swidden cultivation and its associated fallows make a large contribution to the livelihoods of the swidden farmers in the village, mainly by way of self-consumption. In addition, a notable shift was observed in the practice of swidden cultivation from mainly subsistence to a semicommercialized level. There are various possible underlying factors to these changes. The introduction of the government's land use policy in the 1990s, particularly the LFA program, has had a marked effect on the form of swidden agriculture practiced in the village. The LFA program has limited the area available to swidden cultivation, making it difficult for the villagers to pursue their traditional swidden agriculture practice. As a rule, three plots per family are generally allocated for rotational swidden cultivation in villages with high dependence on the practice (Rock 2004). In addition, areas for orchard establishment and horticulture are permitted (Rock ibid). Under the LFA program, the total area available for rotational swidden cultivation in the village was 789.23 ha. The average plot size per household reported by Takeda (2007) was 1.71 ha. In 2014, 124 plots (covering an area of approximately 120 ha) were used

Table 11.3 Animal husbandry in S village in 2011 and 2014

Types of animal	No. of animals	
	2011	2014
Cattle	84	42
Water buffalo	82	91
Goat	161	127
Pig	343	505
Poultry	NA	1093

(*Note* NA—not available)
Source PAFO (2011)

mainly for upland rice, maize, and other crops. Therefore, within the framework of the LFA program, the land for rotational swidden agriculture was still available to a great degree.

In addition, an insufficient level of rice production to meet home consumption was observed in 34 households. According to a report by PAFO (2011), the period for which there was insufficient rice consumption in this village was 2.5 months. Kameda (2015) also reported a rice shortage among swidden farmers in northern Laos. Therefore, it is clearly difficult for the villagers to make a living solely from upland rice cultivation in swidden systems in this area (Kameda 2015).

With the development of cash crop cultivation (especially maize), land use has undergone a great transformation. Some villagers have begun to rent land from neighboring villages, and even within the village. Some households cultivate upland rice for subsistence but cultivate maize in at least one allocated plot to earn cash. This could lead to the establishment of permanent upland fields and thus, might result in various impacts on the environment such as a rise in weeds, erosion, and a negative impact on the livelihood of swidden cultivators in terms of food security and resultant health problems (Wong et al. 2014; Higashi 2015; Kameda 2015). Following limited access to land, the fallow cycles might drop below a critical duration threshold, resulting in negative effects on the productivity of the plots and forest regeneration (Ramakrishnan 1992; Karthik et al. 2009). Consequently, certain ecologists question the sustainability of the practice, since it involves the clearing of primary and secondary forests (Myers 1992; Karthik et al. 2009). Reliance on a single cash crop is risky, as it highly depends on what can be volatile fluctuating market prices. Figure 11.3 showed the unstable trend of market prices for maize. Researchers have pointed out that in the economic valuation of land use in northern Laos, maize and rubber plantation had no potential to contribute to poverty alleviation in the short run (Wong et al. 2014).

The potential of animal husbandry in swidden cultivation has been reported (Hansen 1997) and as a result of development aid programs by NGOs and INGOs, animal husbandry has become increasingly widespread in the area. In 2005, animal husbandry was intensively developed in the village under an international project that provided cattle, buffalo, goats, and pigs for animal husbandry (JICA 2009; PAFO 2011). Thus, there was a marked and significant contribution to villager's livelihood through the rearing of animals. Importantly, out-migration for better job opportunities was observed in this studied village. Due to better access to roads, higher education, and a reduced interest in traditional practices, out-migration will likely increase in the future.

Another important point that requires mentioning is that the implementation of LFA program seems to have increased the extent of secondary forests, and in turn led to an increase in carbon stocks in the fallows, recently promoted by the government of Laos (Hett et al. 2011, 2012). However, a trade-off between food production and carbon stock enhancement is a key issue for REDD+ mechanism in swidden cultivation landscapes (Hett et al. 2011).

Therefore, based on this case study and previous review literature, we need to take into account the importance of swidden agriculture for the livelihood subsistence of swiddeners. Government-sponsored approaches do not work well without the participation of local people as seen in the case study of the Central Highland of Vietnam. Given geographical features do not suit paddy rice fields as with the case study in Palawan, the Philippines. What requires consideration for future planning is an integrated approach including supportive measures such as better access to land, reduced reliance on single cash crops, location-specific conditions for animal husbandry, and the sociocultural context. In addition, as suggested by Wong et al. (2014), swidden cultivation and NTFP collections need to be included in any form of policy incentives under rules and regulations jointly agreed by local communities and government. This study only indicates the livelihood activities of swiddeners, but future research will need to explore the environmental factors of swidden to produce a more complete and nuanced picture.

Conclusion

This chapter has shown that swidden agriculture still plays an important role in the livelihood of Khmu swiddeners in S village, Northern Laos. As noted above, most of the households in this studied village are still engaged in swidden cultivation, not only for their subsistence but also for cash income. According to LFA program, the availability of land was limited, and in turn, this led to the semicommercialization of farming such as maize, black rice, and cassava. Land rental from neighboring villages for cash crop production, especially maize, was also observed. Animal husbandry and NTFPs collection seemed to be the prevailing condition for the sustainability of livelihood of swiddeners.

This chapter has also reviewed the potential impacts of changing swidden in the community. Limits to land access for cultivation could lead to permanent farming, which might result in negative impacts on the environment and the livelihood of swidden practitioners. Also, although sedentary semicommercialized farming in a traditional swidden landscape is the target of the government of Laos, cash crops introduced are highly dependent on the market. On the other hand, an increase in the area of secondary forest and its carbon stock due to the LFA program will need to be maintained.

Under such a situation, the mechanism of REDD+ as promoted by the government, integrated with supportive measures and policy incentives has the potential, as a tool, to conserve secondary forest carbon stocks and to simultaneously promote rural development. However, this means that it will be necessary to further elucidate and understand the role of swidden fallows in carbon stock accumulation, an essential component in any planning of REDD+, not only in Laos but also in other Southeast Asian countries, where traditional swidden practice still exists. In this context, this chapter calls for the importance of an integrated approach that includes local community participation into agricultural planning, especially one that considers the integral role and functions of swidden cultivation.

Acknowledgments This study is a part of collaborative research between Faculty of Forest Science, National University of Laos and Graduate School of Asian and African Area Studies, Kyoto University, focusing on the local use of forest resources in Northern Laos. It was supported financially by Grants-in-Aid from the Ministry of Education, Culture, Sports, Science and Technology, Government of Japan (21255003, 15K01877). We are grateful to the National University of Laos (NUoL) for their cooperation in conducting this collaborative study in Northern Laos, and to the local people of village S for their support and participation. This study would not have been possible without the support of the District Agriculture and Forestry Office (DAFO) and Provincial Agriculture and Forestry Office (PAFO). Special thanks also go to the final-year students of the Faculty of Forest Science, NUoL, Lao PDR, for their assistance during this study.

References

Anoulom, V. (2008). *An ecological study of the swidden fallow vegetation and fallow products of a Khmu village in Nothern Laos*. Unpublished Ph.D. dissertation, Graduate School of Asian and African Area Studies, Kyoto University.

Cramb, R. A. (1993). Shifting cultivation and sustainable agriculture in East Malaysia: A longitudinal case study. *Agricultural Systems, 42,* 209–226.

Cramb, R. A., Colfer, C. P., Dressler, W., Laungaramsri, P., Le, Q., Mulyoutami, E., et al. (2009). Swidden transformations and rural livelihoods in Southeast Asia. *Human Ecology, 37*(3), 323–346.

Colfer, C. J. P. (2008). *The longhouse of the Tarsier, changing landscapes, gender, and wellbeing in Borneo, Borneo Research Council*. ME: Phillips.

Colfer, C. J. P., & Dudley, R. G. (1993). *Shifting cultivators of Indonesia: Managers or marauders of the forests?*. Rome: FAO.

Colfer, C. J. P., Peluso, N. L., & Chin, S. C. (1997). *Beyond slash and burn: Building on indigenous management of Borneo's tropical rain forests*. New York: New York Botanical Gardens.

Dressler, W. (2006). Co-opting conservation: Migrant resource control and access to national park management in the Philippine Uplands. *Development and Change, 37,* 401–426.

Dressler, W., & Turner, S. (2008). The persistence of social differentiation in the Philippine Uplands. *Journal of Development Studies, 44,* 1472–1492.

Ellen, R. F. (2012). Studies of swidden agriculture in Southeast Asia since 1960: an overview and commentary on recent research and sytheses. *Asia Pacific World, 3*(1), 18–38.

FAO/UNESCO. (1979). *Soil map of the world (1:5,000,000): Vol IX Southeast Asia*. Paris: Food and Agriculture Organization of the United Nations and United Nations Educational, Scientific and Cultural Organization.

Fitriana, Y. R. (2008). *Landscape and farming system in transition: case study in Viengkham District, Luang Prabang Province, Lao PDR (Master Thesis)*. Insitut des Regions Chaudes-Supagro Montpelier.

Fox, J., Fujita, Y., Ngidang, D., Peluso, N., Potter, L., Sakuntaladewi, N., et al. (2009). Policies, political-economy, and swidden in Southeast Asia. *Human Ecology, 37*(3), 305–322.

Fujisaki, T. (2012). *Lao PDR REDD + readiness—State of play*. Kanagawa: IGES.

GOL [Government of Laos]. (2007). *Forestry law*. Lao PDR, Vientiane: National Assembly of Laos.

Grandstaff, T. (1980). The transition from shifting cultivation to cash crops: Changes in a land Dayak village. *Studies in Third World Societies, 3,* 113–138.

Hansen, P. K. (1997). Shifting cultivation development in nothern Laos. In E. C. Chapman, B. Bouahom, & P. K. Hansen (Eds.), *Upland farming systems in the Lao PDR—Problems and opportunities for livestock* (pp. 34–42). Canberra: ACIAR.

Heinimann, A., Hett, C., Hurni, K., Messerli, P., Epprecht, M., Jørgensen, L., et al. (2013). Socio-economic perspectives on shifting cultivation landscapes in Northern Laos. *Human Ecology, 41*(1), 51–62.

Hett, C., Heinimann, A., & Messerli, P. (2011). Spatial assessment of carbon stocks of living vegetation at the national level in Lao PDR. *Danish Journal of Geography, 111*(1), 11–26.

Hett, C., Heinimann, A., Epprecht, M., Messerli, P., & Hurni, K. (2012). Carbon pools and poverty peaks in Lao PDR. *Mountain Research and Development, 32*(4), 390–399.

Higashi, S. (2015). An alternative approach to land and forest management in nothern Lao PDR. In Erni, C. (ed.), *Shifting cultivation, livelihood and food security—new and old challenges for indigenous people in Asia* (pp. 253–290). FAO, IWGIA and AIPP.

Hurni, K., Hett, C., Heinimann, A., Messerli, P., & Wiesmann, U. (2013). Dynamics of shifting cultivation landscapes in Northern Lao PDR between 2000 and 2009 based on an analysis of MODIS time series and Landsat Images. *Human Ecology., 41*(1), 21–36.

JICA [Japan International Cooperation Agency]. (2009). *Report on the achievements of the Forest Management and Community Support Project (FORCOM) in Lao PDR.* Ministry of Agriculture and Forestry (MAF) and Japan International Cooperation Agency (JICA), Japan.

Kameda, C. (2015). *Factors influencing the changes of swidden agriculture and its development in rural livelihoods of nothern Laos* Ph.D. Dissertation. Graduate School of Agriculture, Kyoto University, Kyoto, Japan.

Karthik, T., Veeraswami, G., & Samal, P. (2009). Forest recovery following shifting cultivation: an overview of existing research. *Tropical conservation science,* 2(4), 374–387.

Kenney-Lazar, M. n. d. *Shifting cultivation in Laos: Transitions in policy and perspective.* Secretariat of the sector working group for agriculture and rural development (SWG-ARD), Laos.

Kitamura, N. (2004). Forest policy development in Laos. In M. Inoue & H. Isozaki (Eds.), *People and forest—policy and local reality in Southeast Asia, the Russian far east and Japan* (pp. 113–116). Netherlands: Kluwer Academic Publishers.

Laungaramsri, P. (2005). Swidden agriculture in Thailand: Myths, realities, and challenges. *Indigenous affairs,* 2(5), 6–12.

Mertz, O., Padoch, C., Fox, J., Cramb, R. A., Leisz, S. J., Lam, N. T., et al. (2009a). Swidden change in Southeast Asia: Understanding causes and consequences. *Human ecology, 37*(3), 259–264.

Mertz, O., Leisz, S., Heinimann, A., Rerkasem, K., Thiha, Dressler, W., Pham, V., Vu, K., Schmidt-Vogt, D., Colfer, C.P., Epprecht, M., Padoch, C. & Potter, L. 2009b. Who counts? Demography of swidden cultivators in Southeast Asia. *Human ecology,* 37(3), 281–289.

Messerli, P., Heinimann, A., & Epprecht, M. (2009). Finding homogeneity in heterogeneity—A new approach to quantifying landscape mosaics developed for the Lao PDR. *Human Ecology, 37*(3), 291–304.

Myers, N. (1992). Tropical forests: The policy challenge. *Environmentalist, 12*(1), 15–27.

Padoch, C., Coffey, K., Mertz, O., Leisz, S. J., Fox, J., & Wadley, R. L. (2007). The demise of swidden in Southeast Asia? Local realities and regional ambiguities. *Geografisk tidsskrift, 107* (1), 29–41.

PAFO [Provincial Agriculture and Forestry Office]. (2011). *Report on Land-use planning at village level (Samton village) with the aid of AGRISUD International, IRD, NAFRI and CIFOR (in Lao language).* Viengkham: Provincial Agriculture and Forestry Office.

Peluso, N. L. (1996). Fruit trees and family trees in an anthropogenic rainforest: Property rights, ethics of access, and environmental change in Indonesia. *Comparative Studies in Society and History, 38,* 510–548.

Peluso, N. L. (2005). Seeing properties in land use: Local territorializations in West Kalimantan, Indosia. *Geografisk Tidsskrift: Danish Journal of Geography, 105,* 1–6.

Ramakrishnan, P. S. (1992). *Shifting agriculture and sustainable development: An interdisciplinary study from North-Eastern India* (pp. 439–454). Paris: UNESCO.

Rock, F. (2004). *Comparative study on practices and lessons in land use planning and land allocation in Cambodia, Lao PDR*. MRC-GTZ Cooperation Programme, Plascassier, France: Thailand and VIet Nam.

Roder, W. (1997). Slash-and-burn rice systems in transition: Challenges for agricultural development in the Hills of Northern Laos. *Mountain Research and Development, 17*(1), 1–10.

Salemink, O. (2003). *Ethnography of Vietnam's Central highlanders*. London: Routledge.

Schmidt-Vogt, D. 1998. Swidden farming and fallow vegetation in Northern Thailand. Stuttgart: Geoecological Research 8 Franz Steiner Verlag.

Schmidt-Vogt, D., Leisz, S., Mertz, O., Heinimann, A., Thiha, T., Messerli, P., et al. (2009). An assessment of trends in the extent of swidden in Southeast Asia. *Human Ecology., 37*(3), 269–280.

Souvanthong, P. (1995). *Shifting cultivation in Lao PDR—An overview of land use and policy initiatives*. London, UK: IIED.

Spencer, J.E. 1966. *Shifting cultivation of Southeastern Asia.* Berkeley: University of California Press.

Takeda, S. (2013). Commercial cropping and transformation of swidden farming system in a Kham village of Northern Laos (in Japanese). *Japanese Society for Tropical Agriculture, 6*(2), 69–70.

Takeda, S. 2007. Lac cultivation as a strategy for the 'stabilization' of shifting farming in a Khmu village of nothern Laos. In *International Workshop on Sustainable Natural Resources Management of Mountainous Regions in Laos, Louang Namtha Provincial Agriculture and Forestry Office.*

van Vliet, N., Mertz, O., Heinimann, A., Langanke, T., Pascual, U., Schmook, B., et al. (2012). Trends, drivers and impacts of changes in swidden cultivation in tropical forest-agriculture frontiers: A global assessment. *Global Environmental Change, 22*(2), 418–429.

Wong, Y. G., Darachanthara, S., & Soukkhamthat, T. (2014). Economic valuation of land uses in Oudomxay Province, Lao PDR: Can REDD + be effective in maintaining forests? *Land, 3*(3), 1059–1074.

Ziegler, A., Bruun, T., Guardiola-Claramonte, M., Giambelluca, T., Lawrence, D., & Thanh Lam, N. (2009). Environmental consequences of the demise in swidden cultivation in Montane Mainland Southeast Asia: Hydrology and geomorphology. *Human Ecology, 37*(3), 361–373.

Chapter 12
Conclusion: Toward a Research-Based Praxis on Southeast Asia's Ecosystems

Mario Ivan Lopez and Jafar Suryomenggolo

Abstract This conclusion offers an overview of what this volume has engaged in, namely how Southeast Asia's eco-resources have been used and how a research-based praxis can play a role in understanding the challenges its ecosystems face.

Despite a number of shortcomings, the recent United Nations Framework Convention on Climate Change ("The Paris Agreement") offered hope as it was signed, ratified, and adopted by all countries in Southeast Asia.[1] Thailand submitted its Intended Nationally Determined Contribution (INDC) and pledged a 20–30% reduction in its emissions of greenhouse gasses by 2030. Indonesia, one of the world's biggest emitters of greenhouse gasses, plans to protect its tropical rain forest and expand the use of renewable energy, and will raise its target to a 41% cut by 2030 if it receives international aid, such as through REDD+ projects.

Yet, challenges loom large in many other sectors of the region's ecosystems. It's been long documented that Thailand's marine fish resources are overexploited and various reports have suggested that it will need to take serious steps to discipline its fisheries officials in order to fight illegal, unreported, and unregulated fishing activities in the country. Instead of shutting down or putting on hold plans for coal-fired power plants, Indonesia is reorienting its coal reserves to answer the

[1]As of 2013, Southeast Asia as a region contributes around 4.55% of the world's greenhouse gas emissions in which each country, as follows: Brunei: 0.04%; Cambodia: 0.06%; Indonesia 1.64%; Laos: 0.03%; Malaysia 0.67%; Myanmar 0.22%; Philippines: 0.38%; Singapore: 0.12%; Thailand: 0.82%; Vietnam: 0.57%.

M. I. Lopez (✉)
Center for Southeast Asian Studies, Kyoto University, 46 Shimoadachi-Cho Yoshida, Sakyoku, Kyoto, Japan
e-mail: marioivanlopez@cseas.kyoto-u.ac.jp

J. Suryomenggolo
National Graduate Institute for Policy Studies (GRIPS), 7-22-1 Roppongi, Minato-Ku, Tokyo, Japan
e-mail: j-suryomenggolo@grips.ac.jp

M. Lopez and J. Suryomenggolo (eds.), *Environmental Resources Use and Challenges in Contemporary Southeast Asia*, Asia in Transition 7,
https://doi.org/10.1007/978-981-10-8881-0_12

surging national demand for energy and support its infrastructure development. Interestingly, as many as 18 of the country's recent coal power deals (2010–2017) are financially bankrolled, either partly or collectively, by Japan Bank for International Cooperation (JBIC), China Development Bank (CDB), and Korea Development Bank (KDB)—in which the governments of Japan, China, and South Korea are signatories of the Paris Agreement. Running parallel to these issues, for the past decade, the region has witnessed a number of unsettled murder cases of environmental defenders who, in their respective countries, had been advocating eco-friendly initiatives while voicing concerns over environmental degradation and the displacement of local communities due to changes in land use associated with the ongoing capitalist mode of production.[2] Mindful of these and other pressing issues in the context of the changing ecological landscape in the region, governments have worked together under the ASEAN Institutional Framework for Environmental Cooperation. However, a multifaceted ecological pact (*pacte ecologique*) that prioritizes, protects, and conserves its environment is far from reality.

This volume has shown that such path is not impossible if an informed research-based praxis on the region's ecosystem is clearly implemented. Although, far from exhaustive, the breadth of themes and perspectives it has offered are important for field researchers to consider the next step in their contributions and for policy-makers alike to intervene and move toward a common agenda: equitable and sustainable environmental protection. Part one has highlighted the region's geospheric circulation by presenting two important and related issues: the ecological potentials of the region's biodiversity and the bioenergy procurement. These have shown that they are not reducible to a single issue, no matter how large, but rather consist of a complex of challenges that policy-makers need to address without delay.

In part two, this volume has discussed the region's eco-resources governance by focusing on the revitalization of forest resources and the livelihood of forest-dependent communities to go beyond the "political forest" (Peluso and Vandergeest 2001). There are various approaches that can be taken toward a politically viable program of revitalization. In one case, a fair and equitable carbon certificate to measure the potentials of the region's tropical forest may provide a basic income for a local community. The implementation of these have the potential to ameliorate socioeconomic differentiation at the level of municipalities, provinces, and countries, however, they require clear support across multiple levels of governance. In another case, peatland restoration is proven to be the first important step to bring and encourage community participation in sustainable management for the overall health of an ecosystem. These case study-based approaches show that field researchers proactively adopt and implement holistic approaches that include concerns over the economic benefits of forest revitalization for local communities.

[2]To name a few of these cases: Chut Wutty (of Cambodia, in 2012), Sombath Somphone (of Laos, in 2012), Porlajee Rakchongcharoen (of Thailand, in 2014), and Emerito Samarca (of the Philippines, in 2015).

With the chapter on apiculture, we can also see that vulnerability within the region is not limited to human societies, but an issue affecting other key species whose proximity to encroaching urban areas requires a rethinking of how societies share space and maintain high levels of biodiversity.

Part three has discussed the politics of energy and resource use in the region to highlight the contemporary impasse of the region's economic ascent that provides opportunities for sustainable change but at the same time, discredits its biodiversity. The case studies on organic agriculture, dam development, and the transition of swidden agriculture all together show the limits of government's plans on directing the ecosystem, no matter how careful they are formulated. While it does not necessarily mean that human efforts are futile, the protection and conservation of the land, water, energy, and biodiversity will always require a continuous critical assessment and alternative solutions that arise from local needs.

Across Southeast Asia, communities are coming together to create new roadmaps to protect, conserve, and manage the environment. The many ecological issues discussed in this volume and other pressing issues, such as the protection of maritime resources (especially coral reefs), wetlands conservation (such as the mangrove forests of the Straits of Malacca), and the conservation of endangered and threatened species, are indicative of a constant struggle to develop a more inclusive and equitable approach to natural resources governance. We hope this volume will be a stepping stone toward further research and action to better understand these issues on the region's ecosystem.

Reference

Peluso, N. L., & Vandergeest, P. (2001). Genealogies of the political forest and customary rights in Indonesia, Malaysia and Thailand. *The Journal of Asian Studies, 60*(3), 761–812.

Index

M. Lopez and J. Suryomenggolo (eds.), *Environmental Resources Use and Challenges in Contemporary Southeast Asia*, Asia in Transition 7,
https://doi.org/10.1007/978-981-10-8881-0

Printed by Printforce, the Netherlands